The Assay of Spatially Random Material

Mathematics and Its Applications

The Assay of
Spatially Random Material

by

Yakov Ben-Haim

Department of Nuclear Engineering,
Technion-Israel Institute of Technology, Haifa, Israel

Springer-Science+Business Media, B.V.

Library of Congress Cataloging in Publication Data

Ben-Haim, Yakov, 1952–
    The assay of spatially random material.

    (Mathematics and its applications)
    Includes bibliographies and indexes.
    1. Materials– Analysis–Statistical methods. I.   Title.
II. Series: Mathematics and its applications (D. Reidel Publishing
Company)
QD131.B39   1985   543   85–14221

ISBN 978-94-010-8893-0   ISBN 978-94-009-5422-9 (eBook)
DOI 10.1007/ 978-94-009-5422-9

לאשתי מרים
TO MIRIAM

TABLE OF CONTENTS

Chapter 3
Deterministic Design II: General Formulation

Chapter 4
Probabilistic Interpretation of Measurement

Chapter 7
Some Directions For Research

EDITOR'S PREFACE

Growing specialization and diversification have brought a host of monographs and textbooks on increasingly specialized topics. However, the "tree" of knowledge of mathematics and related fields does not grow only by putting forth new branches. It also happens, quite often in fact, that branches which were thought to be completely disparate are suddenly seen to be related.

Further, the kind and level of sophistication of mathematics applied in various sciences has changed drastically in recent years: measure theory is used (non-trivially) in regional and theoretical economics; algebraic geometry interacts with physics; the Minkowsky lemma, coding theory and the structure of water meet one another in packing and covering theory; quantum fields, crystal defects and mathematical programming profit from homotopy theory; Lie algebras are relevant to filtering; and prediction and electrical engineering can use Stein spaces. And in addition to this there are such new emerging subdisciplines as "experimental mathematics", "CFD", "completely integrable systems", "chaos, synergetics and large-scale order", which are almost impossible to fit into the existing classification schemes. They draw upon widely different sections of mathematics. This programme, Mathematics and Its Applications, is devoted to new emerging (sub)disciplines and to such (new) interrelations as exempla gratia:

- a central concept which plays an important role in several different mathematical and/or scientific specialized areas;
- new applications of the results and ideas from one area of scientific endeavour into another;
- influences which the results, problems and concepts of one field of enquiry have and have had on the development of another.

The Mathematics and Its Applications programme tries to make available a careful selection of books which fit the philosophy outlined above. With such books, which are stimulating rather than definitive, intriguing rather than encyclopaedic, we hope to contribute something towards better communication among the practitioners in diversified fields.

Because of the wealth of scholarly research being undertaken in the Soviet Union, Eastern Europe, and Japan, it was decided to devote special attention to the work emanating from these particular regions. Thus it was decided to start three regional series under the umbrella of the main MIA programme.

One characteristic of present day (applied) mathematics is the regular discovery or emergency of new topics to which it can be successfully applied. And the new (from the mathematical point of view) kinds of problems which thus arise, often calling for new ideas of analysis and synthesis. Electrical engineering is one such seemingly inexhaustible source of new interesting problems. Another vast class concerns all kind of problems where knowledge is needed about physically inaccessible spatially distributed structures on the basis of derived signals (such as scattering data or tomographic, i.e. Radon, integrals) to be obtained by the mathematical treatment of the available indirect data. The assay of (radioactive) inhomogeneous materials is one such topic. It generates new problems and calls for new mathematical frameworks, of course not necessarily always immediately of the level of sophistication of more established and older branches of mathematics, but certainly fascinating in the different flavour of the ideas involved and techniques needed. As in many of these new fields, linked directly or indirectly with data processing, the matters of computer algorithms and complexity (as opposed to difficulty) of design play an significant role.

This is a unique book on the design of assay-systems and may well turn out to be a focal point from which one more substantial branch of applied mathematics will flourish. The ingredients, apart from the original problem, include probability, statistics, theory of measurement, control theory and adaptive ideas: a promising mix.

The unreasonable effectiveness of mathematics in science ...

         Eugene Wigner

Well, if you know of a better 'ole, go to it.

         Bruce Bairnsfather

What is now proved was once only imagined.

         William Blake

As long as algebra and geometry proceeded along separate paths, their advance was slow and their applications limited.

But when these sciences joined company they drew from each other fresh vitality and thenceforward marched on at a rapid pace towards perfection.

Joseph Louis Lagrange.

Bussum, July 1985   Michiel Hazewinkel

PREFACE

The assay of material is performed in a multitude of applications. In many situations the structure of the sample is known. By far the most common morphology, occurring almost invariably in the analytical chemistry of solutions, is the homogeneous sample. Despite the importance of such assays, our attention in this volume will be directed to those assay applications in which the structure of the sample is unknown or imprecisely known to the analyst. Not only the distribution of the various components throughout the spatial confines of the sample may be unknown, but also the external shape and dimensions may be hidden. Since the precise morphology of the sample is unknown, one may image that the sample has been drawn from a collection of possible morphologies. It is convenient to refer to this situation as *spatial randomness*, and to describe the assayed material as being *spatially random*. In the introductory Chapter, and indeed throughout the book, we shall discuss numerous examples of spatially random material. From the outset, however, it is important to recognize two points.

First, incomplete knowledge of the structure of the sample causes major complications in the design of a reliable assay system, and in the interpretation of the measurements obtained therefrom. The difficulty in assaying a sample in the absence of precise knowledge of its structure stems from the following fact: different quantities of the material whose assay is to be determined can be differently distributed in the confines of distinct samples, so as to appear identical when assayed. In order to remove this ambiguity from the assay, it is necessary to employ instrumental designs which are totally irrelevant when the sample morphology is known. Furthermore, when the sample morphology is unknown it is likely to be far from adequate to base measurement-calibrations on an assumption of homogeneity (or on any other fixed sample structure).

The second point, which provides the major motivation for

this book, is that a single mathematical framework has been established which underlies the assay of spatially random material.
The reader will have no difficulty accepting this statement as it
pertains to data interpretation, since it will be sufficient to employ standard tools, of acknowledged generality, from the statistical theories of decision and estimation. The case is less clear with
regard to assay-system design, but it would be premature to try
to prove the point in the preface. Let it suffice to say that a general concept of design optimality can be formulated and studied
for a very broad class of spatially random materials, regardless of
whether the specific application is geological, medical, meteorological, hydrodynamical or whatever.

The book has been written with two goals in mind. It is
intended to facilitate the application of a body of mathematical
concepts to the practical tasks of assay-system design and data-
interpretation. A fair portion of the mathematical material, especially that relating to assay-system design, has appeared only in
technical journals which are not readily accessible to workers in the
wide range of disciplines for which this material is relevant. It is
hoped that the large number of examples presented throughout the
book will assist the reader in applying, in his own field of endeavor,
the ideas which are discussed. In light of this aim the examples
are by and large heuristic. Complicated computer calculations are
avoided, as these obscure the concepts which must be implemented
in the course of design or data interpretation. Wherever possible
analytic simplifications are employed so as to enable the reader to
follow in detail the method of analysis. However, the techniques
discussed are by no means limited to simple applications. Great
pains are taken to develop computerizable algorithms by means of
which multi-detector systems of extraordinary complexity may be
efficiently and rationally designed.

The second aim of the book is to engender further study of the
mathematics itself. Consequently, theorems are presented with
proof, especially in the Chapters on assay-system design. The
concept of spatially random material is quite broad, and many

extensions and applications remain unexplored. The final Chapter is devoted to a brief discussion of some directions for further research.

The two aims of the book are to some extent conflicting. The practitioner may view the detailed proofs as extraneous, while the theoretically inclined reader may find the abundance of examples to be distracting. In an attempt to resolve this dilemma, the examples are separated from the mathematical material by appearing in distinct sections. The examples are also fairly self contained. On the other hand, the theoretician should not deprive himself of a good understanding of the problems faced in practice. Similarly, the application-oriented worker is certainly aware that a poor understanding of the concepts underlying a technique may lead to incomplete exploitation or even misuse of the theoretical tools.

Chapter 2 contains the conceptual basis for the design of assay systems. Generalizations and proofs are deferred to Chapter 3. Chapter 4 is devoted to a discussion of the probabilistic interpretation of measurement. In Chapter 5 the statistical tools introduced in Chapter 4 are applied to the topic of assay-system design. Chapter 6 contains an exposition of the technique of adaptive assay, in which the functions of design and interpretation are integrated into a single unit which operates in the course of measurement.

Numerous people have contributed in many ways to the development of this book. Only a few can be mentioned. I am pleased to acknowledge the contributions of Professors Amos Notea and Yitzhak Segal, for it is from their pioneering work on probabilistic nondestructive assay that the present effort has grown. I am deeply indebted to Dr. Natan Shenhav and Professors Ezra Elias and Tsahi Gozani, whose collaboration in various stages of this research was invaluable. Dr. Shenhav's critical reading of a major portion of this book led to the removal of many obscurities.

Yakov Ben-Haim

Haifa
1985

NOTATION

The following symbols are used consistently throughout the book.

SYMBOL	MEANING
X	spatial domain of the sample.
$f(x)$	point-source response function.
$x[h]$	$x_1^{h_1} x_2^{h_2} \cdots x_n^{h_n}$ where h is a vector of non-negative integers.
$r(x)$	source density at point x in the sample.
$F(h)$	point-source response set for the h-moment.
$R(h, u)$	set of all spatial distributions whose h-moment equals u.
$\widetilde{R}(h, u)$	constrained set of allowed spatial distributions whose h-moment equals u.
$C(h, u)$	complete response set for spatial distributions in $R(h, u)$.
$\widetilde{C}(h, u)$	complete response set for spatial distributions in $\widetilde{R}(h, u)$.
$K(Y)$	set of all continuous functions on the set Y.
$E(n)$	n-dimensional real Euclidean vector space.
$E(A)$	statistical expectation of A.

CHAPTER 1

INTRODUCTION

1.1 THE NATURE OF THE PROBLEM

The quantitative assay of material arises in numerous branches of science and industry. In many applications, the assayed sample is heterogeneous, showing variation throughout the sample of density or composition. Such structuring of the sample complicates the calibration of the measurement. It may be feasible to calibrate directly by altering the sample itself, for example by spiking the sample with a known quantity of the analyte. However, for solid samples this may be impossible, or for remote samples such as *in situ* bore-hole geological assays this may be impractical, or for medical assays on living human subjects this may be undesirable for health reasons. On the other hand, preparation of standardized samples which simulate the spatial structure of the assayed sample may be technically quite difficult. A more serious problem arises when the spatial structure of the sample is unknown, such as for a subterranean mineral deposit. Determination of the spatial structure of the sample may be much more challenging than the assay itself. Finally, the spatial structure may vary randomly throughout an ensemble of samples. Random variation of the spatial structure is observed in the assay of fissile materials produced in various stages of the nuclear fuel cycle, in parameter-estimation of flowing fluids, in medical assays and in geological and meteorological studies.

The analyst wishing to assay a material with random or unknown spatial distribution faces two tasks. First, he must design the assay system so as to reduce the uncertainty of the assay. Second, he must interpret his measurements so as to accurately represent the true content of the sample. These two tasks — de-

1

sign and interpretation — are complementary aspects of the assay of spatially random material, and together constitute the theme of this book.

The sources of assay-uncertainty are two-fold. There is *statistical uncertainty* arising from randomness of the measurement process. Statistical uncertainty is not unique to the assay of spatially random materials. It is generally well understood and usually readily evaluated. For instance, the randomness of many discrete counting processes is well described by Poisson statistics. The second source of uncertainty is the random or unknown spatial distribution of the analyte [1]. Depending on how the analyte is distributed in the sample, a large or small count rate may be sensed by the detector. Stated conversely, different quantities of analyte, arranged in different spatial distributions in their respective samples, may give rise to the same count rate in the detector.

The instrumental basis for reducing this *spatial uncertainty* has reached a high degree of sophistication. The development of very sensitive detectors for penetrating radiations of many types has greatly expanded the capability of the analyst. On the other hand, the theoretical ramifications of spatial uncertainty have never been analyzed in a systematic way nor been given a unified treatment. Mathematical techniques for rigorously assessing the spatial uncertainty of a proposed assay-system design have, until recently, been lacking. As a result, the design of assay systems which are intended to measure spatially random material has relied very heavily on costly and time consuming trial and error.

The array of conceivable and technically feasible assay-system designs which are suitable to any given application is vast. The decisions confronted by the designer may be grouped into the following categories.

1. Passive or active assay. If the analyte spontaneously emits a measurable radiation, the designer may choose to passively detect that radiation. For example in the assay of *in situ* uranium ore it is common to measure the 1.76 MeV gamma radiation of the daughter isotope ^{214}Bi produced in the decay chain of ^{238}U. How-

ever if geochemical or other evidence indicate that the uranium is not in equilibrium with its daughters, due perhaps to leaching processes, it may be necessary to measure the uranium directly. This can be done, for example, by neutron activation. Indeed, in many assay applications it is advantageous to activate the sample with an external source of radiation. The choice of whether to measure actively or passively is one of the most basic decisions facing the designer.

2. Type of radiation. Closely related to the type of assay is the choice of the radiation with which to activate the sample and with which to detect the analyte. Electromagnetic radiation, neutrons, electrons and other types of radiations at a wide range of discrete energies or energy bands are all available. Each energy and type of radiation presents special features with regard to transport properties, reactions with the matrix or the analyte, efficiency of detection and so on.

3. Type of detector. Inseparable from the choice of the radiation is the choice of the detector. Detectors vary according to efficiency and speed of detection, energy resolution, cost, durability and portability.

4. Number of detectors. Very appreciable improvement in overcoming spatial uncertainty can be attained in many assay applications by using more than one detector or by repeating the measurement at various points around the sample. Since the cost and complexity of the assay increase with the number of detectors or measurements, the designer must make a careful choice on this issue.

5. Deployment of detectors. Proper choice of the number of detectors or number of measurements requires consideration of their geometrical deployment around the sample. The power of multiple measurement can be lost by unwise placement of the detectors. In some assay applications it is advisable to have either the sample or the detector move during measurement to achieve a certain degree of spatial averaging. This is common in some nuclear safeguards assays as well as in certain medical assays. The

detector or sample movement can also be done adaptively, by on-line feedback control.

6. Measurement duration. It will be seen (in Chapter 2) that statistical and spatial uncertainty present the designer with conflicting interests. With the intention of reducing the spatial uncertainty, the designer will usually tend to favor detector deployments in which the detectors are far from the sample. This makes the sample more "point-like" and thus diminishes the spatial uncertainty. However displacement of the detectors from the sample will usually reduce the count rate and increase the statistical uncertainty. The total amount of time which can be devoted to the measurement of each sample is an independent factor of primary importance in determining the degree of statistical uncertainty. If the designer is free to specify a lengthy measurement duration, he will be able to choose a detector deployment which is specially suited to diminishing spatial uncertainty. On the other hand, if strict time-constraints are imposed on the measurement, the assay-system design may take a radically different form.

7. Size and shape of the sample. The designer may be able to specify the size or shape of the sample. Choice of a small sample size may reduce the spatial uncertainty. However, a greater number of samples may thereby need to be analyzed. If a time constraint is imposed, the increased statistical uncertainty of each measurement may outweigh the improvement of the spatial uncertainty.

8. Matrix composition. In some applications the designer may be free to choose the density or composition of the matrix of the sample. This may influence the radiation-transport characteristics of the sample and thereby be of importance in the design process.

From this brief survey it is plain that the array of design decisions is complex and highly interconnected. It is clearly important for the designer to have an efficient and rigorous means of evaluating the statistical and spatial uncertainty of any proposed design. The evaluation should be based on a quantitative measure of performance which enables the designer to select the design best suited to his application. The measure of performance should be sufficiently comprehensive so as to provide realistic representation of each design. On the other hand the data base and computational effort required for evaluating the performance should be modest enough to allow ready examination of a large number of broadly different design concepts.

But careful design is not enough. The analyst must be able to properly interpret the results of his measurement. Statistical and spatial uncertainty result from stochastic processes: randomness of the measurement and randomness or uncertainty of the spatial distribution of the analyte. The analyst must exploit his knowledge of these processes so as to maximize the reliability of his interpretation of the measurements performed.

Assay-system design and measurement-interpretation are complementary tasks, and should be treated in a unified approach. The techniques and concepts developed for one aspect of this subject often prove useful for the other. For example, the probabilistic formulation of assay interpretation is fundamental to the probabilistic techniques for design. In the most advanced assay systems, design and interpretation are linked in an adaptive assay process: the measurements are interpreted on-line and are fed back to the assay system to cause adaptive modification of the assay-system design. Adaptive assay is of vital importance in any application in which *a priori* knowledge of the sample is limited. The techniques of adaptive assay result from a thorough synthesis of the processes of design and interpretation.

1.2 EXAMPLES OF SPATIALLY RANDOM MATERIAL

Random spatial distributions of material arise in a wide range of applications and in surprisingly diverse forms. In order to appreciate the generality of the problem, it is worthwhile to mention a selection of examples.

1. Nuclear Safeguards

Safe operation of the nuclear fuel cycle requires the accurate assay of small quantities of fissile material appearing in various types of samples. The unifying features of these assay problems are that the spatial distribution of the fissile material varies randomly from sample to sample, and that the sample is unavailable for direct inspection because of radiological hazards.

(a) Low-level process waste. The samples are randomly filled and hermetically sealed containers carrying radioactive waste from chemical processing of fissile fuel. The analyte occurs as a small number of discrete fragments of fissile material [2].

(b) Leached fuel hulls. Nuclear reactor fuel is typically encased in long thin tube-like cladding. After irradiation in the reactor core, the fuel rods may be chopped into short segments from which the spent fuel is removed by the leaching action of an acid solvent. A small quantity of fissile material may occasionally remain on the inner surface of a leached fuel hull. These hulls are assayed in disordered clusters for detection of this fissile material, which is present in a highly random spatial distribution [3].

(c) Spent-fuel. In some situations the irradiated rod must be assayed for fissile material before chemical processing. Assay of fissile material remaining in an irradiated fuel rod is complicated by an imprecisely known and variable distribution of fissile material along the length of the rod [4].

(d) Separation column. In the chemical processing of dissolved irradiated fuel, the fissile material may be isolated in a counter-current chemical separation column. The radial and axial distribution is random to a certain degree, or unknown with any great precision, which complicates the on-line assay of the fissile material in the column.

2. Medical Assays

In many medical assays of *in-vivo* radionuclides one confronts an unknown spatial distribution of the assayed material. The ease of non-intrusive assay of a trace quantity of a radioactive isotope makes this an attractive alternative to standard chemical analysis. However the analyst must overcome the problems presented by unknown or random spatial distribution of the isotope. For radiological reasons, usually only passive assay is feasible.

(a) Assay of naturally occurring or unintentionally introduced isotopes. The assay of ^{40}K is a common muscle-mass diagnostic tool. Calibration with simulated models of the spatial distribution in the body is quite accurate for a population of healthy individuals [5], but may be of limited accuracy if the measured individuals display pathological deviation from the norm. Other radionuclides which may be found in the body include fallout fission products such as ^{137}Cs, ^{90}Sr [6] and ^{131}I [7], which enter the body through the food chain, and various nuclides accidentally taken in by workers in the nuclear industry [8].

(b) Assay of intentionally introduced isotopes. When radioisotopes are introduced into the body intentionally for research or diagnostic purposes, it is possible to obtain whole-body burdens indirectly by collecting all the excreta and measuring its radioactivity. However, in these cases whole-body counting offers an attractive alternative, simplifying the measurement and avoiding errors due to incomplete collection of excreta. Radionuclides which are used in this way include: 1) ^{59}Fe as ferrous citrate, ^{51}Cr-labelled red blood cells and ^{57}Co-labelled vitamin B_{12} for hematological studies [9]. 2) Various radioactive metals which are used in the diagnosis of trace element deficiencies. These include ^{67}Cu [10] for patients with the Wilson disease, ^{65}Zn [11] for zinc deficiency related diseases and ^{51}Cr for glucose metabolism diseases. Another important application of whole-body counting is for measuring the retention of new drugs by radioactive labelling. Here the toxicity of the drug is unknown and only very small quantities can be used in testing with human volunteers. In addition, calibration with phantoms is unfeasible since the distribution of the radionuclide in the body in many cases is unknown and varying in time. All these assay applications require assay systems which are insensitive to the variable or unknown time-dependent spatial distribution of the drug in the body [12].

(c) Assay of particles in the lungs. In some situations it is necessary to assay small aerosol particles deposited in the lungs. The measurement is quite sensitive to where the particles happen

to be deposited. This is an acute problem when the assay is based
on measurement of low-energy radiation emitted by the particles,
which is strongly absorbed in the lung tissue [13].

3. Mineral Evaluation

Assay techniques based on nondestructive measurement of radiation are readily automated and adapted to remote operation.

(a) Uranium prospecting. Natural uranium deposits in the
earth are evaluated with standard radiation-detection instrumentation lowered down a core drilling [14]. In certain geomorphological conditions the uranium appears as small randomly dispersed
nodules. Accurate assessment of the richness of the deposit requires consideration of the spatial randomness of the source material.

(b) Coal assay. In a coal-processing plant the composition of
coal (sulfur content, ash content, etc.) may be evaluated while the
coal moves on a conveyor belt. The assay must be designed to be
insensitive to the variable size, shape or density of the overall coal
sample and of the individual coal chunks [15].

(c) Particle-Size Effects. Measurements of the intensity of
fluorescent X-rays [16] or of the transmission of gamma radiation
[17] are used in many quantitative measurements of density and
composition of ores, alloys, industrial slurries or other composite
materials. Calibration of such measurements is complicated by
porosity or granularity of the sample or by variable size, shape
and spatial distribution of the component materials.

4. Measurement of Fluid Flow and Related Applications

The assay problems discussed up to now have concentrated on
measuring the total amount of the assayed material. This may
be viewed as measurement of the zeroth moment of the spatial
distribution of the assayed material. In some situations one wishes
to measure higher moments of the distribution. The techniques
for design and interpretation of a zeroth-moment assay can be
immediately extended to first- and higher-moment assays.

(a) Flow rate. A standard technique for measurement of the
flow rate of a fluid involves injection (or activation) of an isotope

at one point in the flow channel, and measurement of the time at which the isotope passes a different point downstream [18]. Due to dispersive effects in the flowing fluid, the isotope passes the detection point as a spatial distribution whose shape is often not well known. Thus the problem of determining when the isotope passes the detection point requires measuring where the material is at a sequence of points in time. This measurement is in fact the assay of the first moment (average position) of the unknown spatial distribution of the isotope.

(b) Bubble detection. The detection of the presence and location of a bubble in otherwise single-phase flow may be approached as a zeroth- and first-moment assay problem.

(c) Flow regime determination. As a generalization of the bubble-detection problem, the design of an assay system for evaluating the flow regime in multi-phase flow may be approached as a low-order-moment assay. For example, distinguishing between plug flow, annular flow and inverted-annular flow can be based on an assay of the first spatial moment of one of the phases.

(d) Turbulence estimation. The evaluation of turbulence in a flow channel may be done by evaluating the degree of spatial dispersion of a point-injection of isotope. This involves a second-moment assay.

(e) Moisture content of soil. In agricultural and other applications it is necessary to assess the moisture content of soil. This is complicated if the structure, composition or density of the soil is variable or unknown. The moisture content can be measured by assaying low-order moments of the spatial distribution of the slow-neutron flux generated by a neutron source [19].

From these examples it is apparent that by formulating a general approach to the problems of design and interpretation, which arise in connection with random or unknown spatial distributions, one is able to handle a very broad range of assay applications. We have identified problems of spatial randomness in zeroth-moment assays in nuclear engineering, medicine, geology and other fields. Spatial randomness arises in first- and second-moment assays in various measurements of fluid flow and related areas. The extension from zeroth-moment assay (the usual type of assay) to first- and second-moment assay will be justified by establishing the fundamental mathematical unity of these measurement problems.

However, care should be taken not to carry this extension too far. The assay of low-order moments of an unknown or random spatial distribution is designed to yield precisely defined yet limited information about the spatial structure of the analyte. Only in very special circumstances can such an assay completely characterize the spatial distribution of the analyte. The general problem of reconstructing a spatial distribution is in the province of tomography. The theoretical basis of tomographic imaging is adapted to the exacting demands of image reconstruction, and the tomographical problems of large dimensionality and of ill-conditioned transformation are usually irrelevant for the assay of low-order moments. Consequently, it would be unproductive to apply a tomographical approach to the design and interpretation of low-order-moment assays, and a different approach is called for.

The material of Chapters 2 through 4 forms the theoretical infrastructure for the assay of low-order moments of a random or unknown spatial distribution. In these Chapters we discuss and interrelate various concepts from the theory of probability, the theory of detection, estimation and decision, and the algebraic and geometrical theory of convex sets. The main thrust is how these ideas are molded together to give a unified approach to assay-system design and measurement-interpretation for the broad range of applications in which spatial randomness arises.

Chapters 2 and 3 discuss a deterministic measure of performance for evaluating assay-system designs. The central concept of these Chapters is the idea of relative mass resolution, defined as follows. Consider a quantity m of assayed material. Let $m + d$ be the smallest quantity greater than m such that any spatial distribution of m grams is distinguishable from any spatial distribution of $m + d$ grams. Then $(m + d)/m$ is the relative mass resolution. The concept of relative mass resolution is a useful and meaningful measure of performance for zeroth-moment assays, enabling quantitative comparison of alternative proposed assay-system designs. This measure is deterministic in the sense that any spatial distribution of m grams is distinguishable from any spatial distribution of $m + d$ grams. Probabilistic generalizations, which account for the relative probability of different spatial distributions, are deferred to Chapter 5.

For single-detector systems it is usually fairly simple to evaluate the relative mass resolution. However many well known assay problems require the use of multiple-detector arrangements in order to achieve satisfactory mass resolution. The bulk of Chapter 2 is devoted to developing an efficient computerizable algorithm for evaluating the relative mass resolution for multi-detector configurations. After establishing this algorithm for zeroth-moment assay, its generalization to the assay of other moments of the spatial distribution is developed. Additionally, the algorithm is extended to include the statistical uncertainty of the measurement. The mathematical basis of the algorithm involves concepts not usually encountered in detection and estimation problems. The detailed discussion of the mathematical foundations, in conjunction with numerous examples, will enable the reader to apply the algorithm to his own assay problems.

Chapter 4 turns to the problem of probabilistically interpreting the results of measurement. The mathematical material of this Chapter is not new, and it is assumed that the reader has a reasonable background in the standard tools of probability. The major aims of this Chapter are two-fold. First, we wish to present the concepts of decision and estimation theory in a formulation relevant to the assay of spatially random materials. Second, we wish to establish those basic tools and concepts which, in the following Chapter, will be employed to develop probabilistic generalizations of the measure of performance discussed in Chapters 2 and 3. We begin by dealing with the probability density of the measurement, our aim being to show how one actually evaluates this probability density for complex multi-detector systems. Then we present the decision theories of Bayes and of Neyman and Pearson. Finally we consider the evaluation of probabilistic calibration curves.

Chapter 5 returns to the design of assay systems. The aim of this Chapter is to develop measures of performance which incorporate the relative probability of different spatial distributions of the assayed material. These performance-measures are tools for design-decision just as is the deterministic measure developed in Chapters 2 and 3. The motivation for the probabilistic measures is as follows. We will find from examples in previous Chapters that for many assay problems the deterministic mass resolution is quite large. In fact in some cases of practical interest it is even infinite. This arises from the fact that very unlikely spatial distributions are considered along with more probable spatial distributions. In order to overcome this difficulty it is necessary to consider the probability of different spatial distributions. This can be done in a variety of ways, each leading to a different degree of comprehensiveness and computational difficulty of the measure of performance. The simplest is the relative error criterion, which does not require explicit evaluation of the probability density of the measurement. The next performance-measure is based on the concept of minimum variance, and employs the Rao-Cramer inequality. The final and most complex measure is a probabilistic generalization of the deterministic measure of performance developed in Chapters 2 and 3.

In our analysis of the assay of spatially random materials we have separated the problem of assay-system design from the problem of interpreting the measurements obtained from a given assay system. We have however exploited the fact that techniques for probabilistic interpretation of measurement are extremely useful for design decisions. In Chapter 6 we go one step further and discuss assay systems which adjust their own design parameters in the course of measurement. This *adaptive assay* is particularly useful in assay problems for which limited information is available, before measurement, on the stochastic properties of the sample. We begin by discussing assay algorithms for adaptive — on-line

— determination of the duration of measurement by the technique of sequential analysis. We then consider assay algorithms which adaptively choose the positioning of the detectors around the sample. Three different applications are studied. Each of these three examples employs a different measure of performance. The possible approaches to these problems are numerous and diverse, and our treatment is not intended to be exhaustive. Rather, the emphasis is on illustrating the utility of the techniques developed earlier in this book.

Chapter 7 is devoted to a consideration of some important advanced topics not covered in the book, and which constitute avenues for research today. First we consider problems arising from nonlinearity of the detector response to analyte material at high density. Next we consider the problems arising when the response set (to be defined in Chapter 2) is non-convex. For our next topic we turn to the problem of asymptotic designs: what is the resolution obtainable from an infinite number of detectors? Can this resolution be evaluted directly, without need for evaluating the resolution with a finite number of detectors? If so, such a technique can provide a direct indication of the best possible design. Such a tool would be very useful in preliminary comparison of broadly different design concepts. In the next section we consider the design of an algorithm for the isolation of malfunctions in a complicated dynamical system. The technique for assay-system design can be applied to optimize the algorithm even though this measurement is not a material assay problem. Finally, we discuss some advanced concepts which may be applied to the problem of adaptive assay systems. Building on and extending the mathematical foundations developed in the book, it is hoped that solutions to these and other problems will be forthcoming.

NOTES

[1] We shall use the term *analyte* to refer to the specific material within a sample whose assay is to be determined. The analyte is distinguished from the remaining constituents of the sample,

which are called the *matrix*. The analyte is sometimes referred to
as the *source material*.

[2] 1. R. M. Auguston and T. D. Reilly, Fundamentals for Passive
Non-destructive Assay of Fissionable Material, Los Alamos Scientific Laboratory report no. LA-5651-M, 1974.
2. A. Notea, Y. Segal, A. Bar-Ilan, A. Knoll and N. Shenhav,
Proc. Int. Meeting on Monitoring Pu-Contaminated Waste, Ispra, Italy, 1979.
3. A. Knoll, A. Notea and Y. Segal, Probablistic Interpretation of
Nuclear Waste Assay by Passive Gamma Technique, *Nucl. Tech.*,
56: 351 (1982).
4. R. A. Harlan *et al*, Operational Assay for Fissile Material in
Crated Nuclear Energy Wastes, *6-th Symp. on Safeguards and
Nuclear Material Mgt.*, ESARDA, Venice, May 1984.

[3]   G. G. Slaughter *et al*, A Proof-of-Principle Experiment for
Nondestructive Assay of Leached Fuel Hulls, Oak Ridge Natl. Laboratory report no. ORNL/TM-8445, (1983).

[4]   M. S. Zucker *et al*, A Coincidence Technique to Reduce Geometry and Matrix Effects in Assay, ESARDA, *5-th Ann. Conf.
on Safeguards and Nuclear Material Management*, pp 277-85, Ver-
sailles, 1983.

[5] R. A. Dudley and A. Ben-Haim, Comparison of Techniques
for Whole-Body Counting of Gamma-Ray Emitting Nuclides with
NaI(Tl) Detectors, *Phys. Med. Biol.*, 13:181-93, 194-204 (1968).

[6]   R. A. Dudley and A. Ben-Haim, Assay of Skeletally Deposited Strontium-90 in Humans by Measurement of Bremsstrahlung,
Health Physics, 14: 449-59 (1968).

[7]   J. Lando and D. Newton, Some Recent Measurements of ^{137}Cs
and ^{95}Zn in Human Beings, *Nature*, 195: 851 (1962).

[8]   W. W. Parkinson jr., R. E. Goans and W. M. Good, Realistic
Calibration of Whole-Body Counters for Measuring Plutonium, in
*National and International Standardization of Radiation Dosime-
try*, IAEA Conf., Vienna, Vol., 2, pp 155-77 (1977).

[9] M. F. Cottral, M. E. A. O'Connell, M. G. Trott and D. G. Well, Some Applications of Whole Body Scanning in Haematology, *Brit. J. Haematology*, 25: 545-6 (1974).

[10] W. N. Tauxe *et al*, Body Retention of Injected ^{69}Cu ... Comparison of Whole-Body Counting Systems, *Mayo Clinic Proc.*, 49: 382-6 (1974).

[11] B. E. Arvidson, A. Coderld, E. B. Rasmussen and B. Sandstrom, A Radionuclide Technique for Studies of Zinc Absorption in Man, *Int. J. Nucl. Med. Biol.*, 6: 104-9 (1978).

[12] S. H. Cohn *et al*, Recent Advances in Whole-Body Counting: A Review, *J. Nucl. Med. Biol.*, 1: 155-65 (1974).

[13] F. A. Fry *et al*, A Realistic Chest Phantom for the Assessment of Low Energy Emitters in Human Lungs, *4-th Intl. Conf. of Intl. Rad. Prot. Soc.*, vol. 2, pp 475-8, Paris, 1977.

[14] International Atomic Energy Agency, Radiometric Reporting Methods and Calibration in Uranium Exploration, *Technical Report Series*, No. 174, Vienna, 1976.

[15] E. Elias and T. Gozani, Nuclear Assay of Coal, Vol. 2: *Coal Composition Determination by Prompt Neutron Activation Analysis — Theoretical Modeling.* EPRI report FP-989 vol. 2, April 1980.

[16] 1. R. P. Gardner, D. Betel and K. Verghese, X-Ray Fluorescence Analysis of Heterogeneous Material: Effects of Geometry and Secondary Fluorescence, *Int. J. Appl. Rad. Isot.*, 24: 135-46 (1973).
2. F. Claisse and C. Samson, Heterogeneity Effects in X-Ray Analysis, *Adv. X-Ray Analysis*, 5: 335-54 (1962).
3. A. Lubecki, B. Holynska and M. Wasilewska, Grain Size Effect in Non-Dispersive X-Ray Fluorescence Analysis, *Spectrochimica Acta*, 23B: 465-79 (1968).

[17] 1. H. L. The, W. Michaelis and H.-U. Fanger, Partice Size Effects in Non-Destructive Material Assay by Gamma Ray Absorptiometry, *Nucl. Instr. Meth.*, 212: 445-61 (1983).

2. K. Umiatowski *et al*, Gamma Ray Absorption Coefficient for Heterogeneous Materials, *Nucl. Instr. Meth.*, 347-51 (1977).

[18] 1. P. Kehler, Accuracy of Two-Phase Flow Measurement by Pulsed Neutron Activation Techniques, in *Multiphase Transport Fundamentals, Reactor Safety Applications*, Vol. 5, p 2483, Hemisphere Pub., 1980.
2. M. Perez-Griffo, R. C. Black and R. T. Lahey, *Basic Two-Phase Flow Measurements Using* ^{16}N *Tagging Techniques*, NUREG/CR-0014, Vol. 2, p 923, 1980.
3. P. B. Barrett, An Examination of the Pulsed-Neutron Activation Technique for Fluid Flow Measurements, *Nucl. Eng. Design*, 74: 183-92 (1982).

[19] D. Ingman and E. Taviv, Development of the Moments Method for Neutron Gauging, *Nucl. Instr. Meth.*, 190: 423-31 (1981).

CHAPTER 2

DETERMINISTIC DESIGN I:
CONCEPTUAL FORMULATION

2.1 RELATIVE MASS RESOLUTION

In this Chapter we shall develop a powerful analytical aid for
the design of assay systems which are intended to measure materials which are randomly distributed in space. We learned in
Chapter 1 that the design procedure must be based on a quantitative measure of performance which guides the designer to the
best from among alternative realizable designs. This performance-
measure must evaluate the accuracy of the assay system in the
face of two main challenges. First of all, the measure must assess
how well the assay system confronts the challenge of spatial uncertainty: the spatial randomness of the assayed material. Second,
the measure of performance must evaluate the degree to which
the assay system is sensitive to the statistical uncertainty arising
from randomness of the radiation-measurement process. Not only
must the measure encompass these diverse and complex aspects
of an arbitrary assay system, it must also be practicable. That is,
it must be computationally feasible to evaluate the performance
measure for a wide range of alternative proposed designs.

The central concept of this Chapter is the elementary idea of
relative mass resolution, defined as follows: Consider a quantity m
of source material. Let $m+d$ be the smallest quantity greater than
m such that any spatial distribution of m grams is distinguishable
from any spatial distribution of more than $m + d$ grams. Then
$(m+d)/m$ is the relative mass resolution. This concept of relative
mass resolution is a useful and meaningful criterion for evaluating the performance with regard to the spatial uncertainty, by
enabling quantitative comparison of alternative proposed assay-

system designs [1]. Towards the end of this Chapter we will find it to be an easy matter to extend the concept to include evaluation of statistical uncertainty as well.

The relative mass resolution is a deterministic measure of performance in the sense that *any* spatial distribution of m grams is distinguishable from any spatial distribution of more than $m + d$ grams. We shall find that the deterministic nature of this criterion is responsible for usually providing a conservative estimate of the resolution capability of the assay system. After our study in Chapter 4 of techniques for probabilistic interpretation of measurement, we will be ready to discuss probabilistic generalizations of the deterministic measure of performance. The utility of the deterministic criterion remains, despite its limitations, in its ease of application to a large range of proposed assay-system designs, and in its succinct assessment of assay capability. The deterministic performance measure facilitates a first-stage design analysis whose aim is to narrow the field to a limited number of design alternatives.

For single-detector systems it is usually fairly simple to evaluate the relative mass resolution. However, many well known assay problems require the use of multiple-detector arrangements in order to achieve satisfactory mass resolution. The bulk of this Chapter is devoted to developing an efficient computerizable algorithm for evaluating the relative mass resolution for multi-detector configurations.

2.2 RESPONSE FUNCTIONS

The fundamental building block for all our subsequent work is the point source response function of the detector, which we denote $f(x, y)$. The response function specifies the time-averaged count rate produced by a detector located at position y in response to one unit of the analyte concentrated as a point-source at position x in the sample [2]. The rest of the assayed sample is assumed to consist of non-emitting matrix material of known spatial structure. Point sources are idealizations of reality, and in practice we measure the response function with a source whose dimensions are small compared to the mean free paths of the activating and detected radiations [3]. In the most general case, x and y are both three-dimensional vectors. In many practical situations considerations of symmetry reduce the dimensionality of x and/or y. The response function may include the effects of absorption and scatter of the detected radiation in the matrix and in the detector, geometrical attenuation and efficiency of the detector. For assays based on activation of the assayed material by an external source of radiation, the response function includes the various factors governing the efficiency of this activation.

Throughout our study we shall assume two very important linearity properties of the response function. The first is *linearity in time*: The number of counts to be expected in duration t is precisely t times the time-averaged response.

This assumption will be valid if the activity of the sample is constant, or at least very nearly constant, for the duration of the measurement. We will not encounter any important practical situations in which this is not the case.

The second linearity assumption is *linearity in mass*. It can be expressed in either discrete or continuous forms. The discrete form is:

(i) $\sum r(i) f\big(x(i), y\big)$ is the time-averaged count rate in response to sources of intensity $r(i)$ embedded in the matrix of the sample at positions $x(i)$.

The continuous version of this assumption is:

(ii) $\int r(x) f(x, y) dx$ is the time-averaged count rate in response to a local source density $r(x)$ embedded in the matrix of the sample.

The assumption of linearity in mass states that the responses to different source deposits are linearly superimposed at the output of the detector. This assumption will be valid if the concentration of source material is sufficiently low so as to not appreciably affect the radiation absorption characteristics of the sample. This however does not require that the absorption within the deposits of source material be the same as in the matrix. In the case of discrete source deposits of significantly different absorptivity from the matrix, linearity in mass will hold if the number density of the deposits is low enough so that "shadowing" of one deposit by another occurs with very low probability [4]. Alternatively, if the absorption characteristics of the source material are the same as those of the matrix, then the assumption of linearity in mass will be valid even at high source concentration.

The response function is the operational manifestation of the assay-system design characteristics. The response function derives its properties from the type of detector and its mode of operation, from the nature of the source and matrix materials, and from the types of radiations employed. Response functions of very different sorts are routinely encountered, and few generalizations characterize them all. For passive gamma assay systems the response function of the unscattered radiation is almost invariably a monotonically decreasing function of the source-detector distance. However, when a wide energy-window is measured, the effects of radiation buildup may cause the response to increase with increasing source-detector distance for a certain range of distances. For active delayed-neutron assay the response function is not necessarily decreasing or even monotonic at all [5]. In some applications the response function may vary by orders of magnitude as the point source is moved through the sample, while in other cases the response function may be nearly independent of the source position. It is evident that we must pursue a rather general course if we hope to develop a measure of performance which will encompass the entire spectrum of realistic assay-system design options.

2.3 POINT-SOURCE RESPONSE SETS

We very often encounter assay systems comprising several detectors located at positions $y_1, \ldots, y_n$. Each detector is characterized by its own response function, denoted by $f(x, y_i)$. When the number and position of the detectors are fixed, it is convenient to represent all the response functions together in a single *vector response function* defined by

$$f(x) = \big(f(x, y_1), \ldots, f(x, y_n)\big) \qquad . \qquad (2.3.1)$$

This vector represents the set of time-averaged count rates produced by the n detectors in response to a single point source at position x, when the rest of the assayed sample consists of non-emitting material of known structure. The linearity properties of the individual response functions hold also for the vector response function.

By employing the vector response function $f(x)$ we avoid the unnecessary burden of continually referring to the detector locations. In fact we usually have no particular reason to keep track of the point-source position either; all that really is of interest is the set of vector responses. Consequently we shall define the *point-source response set* as the set of all n-vector values which the vector response function may assume. In order to facilitate a formal definition of the point-source response set, let X represent the set of all positions x within the sample at which the point source may be located. Then the point-source response set is defined as the set

$$F = \{f : f = f(x) \quad \text{for all} \quad x \in X\} \qquad (2.3.2)$$

Let us consider some elementary properties of the sets X and F. They are both subsets of real Euclidean spaces. X will often be three-dimensional, for example in the assay of *in-vivo* radionuclides or of nuclear waste containers. X may be a two-dimensional set, for example in the assay of surface deposits. In some cases X may be treated as a one-dimensional set; this may arise for example when the detector-sample geometry has a sufficiently high degree of symmetry. The dimensionality of F, the point-source response set, equals the number of detectors in the assay system. In practice both X and F will be closed sets [6]. This is a convenient fact which will simplify our notation later, though it is not of fundamental importance.

In medical, thermohydraulic and nuclear engineering applications one is accustomed to encounter samples of finite size, which means that X is a bounded set. In geological prospecting the dimensions of the sample may be much greater than the mean free paths of the relevant radiations, and thus effectively infinite. In such a case X is an unbounded set. The point-source response set F may or may not be bounded, regardless of whether or not X is bounded. For example, the response of a single gamma-radiation detector can be made instrumentally unmeasurably large by allowing the detector-source distance to approach zero, irrespective of the physical extent of the sample. Conversely, if the detector is removed somewhat from the sample surface, the response function is bounded, even for a very extensive sample. A further very important point is that when X is unbounded, F will be likely to contain the origin. This will be seen to be a special case of an important geometrical property of the set F which will cause peculiar and challenging difficulties, especially in geological applications, and which will provide one of the motivations for our study of probabilistic design analysis in Chapter 5.

Finally we note that X will often, though not invariably, be a convex set [7]. On the other hand F will very often not be convex. Consider for example a cylindrical sample, with height appreciably greater than diameter. Let us place a passive gamma detector at each end of the sample on the rotational symmetry axis, as in figure 2.3.1. The point-source response set will be something like that shown in figure 2.3.2. When the point-source is near the top of the sample, the count rate in detector 1 will be high and in detector 2 it will be low. The situation is reversed when the point source is near the bottom of the sample. The crescent-like appearance of the response set arises from the strongly non-linear dependence of the response function on the detector-source separation. The thickness of the crescent arises from the non-zero width of the sample. In later sections we will pursue a more thorough study of several realistic examples.

Detector 1

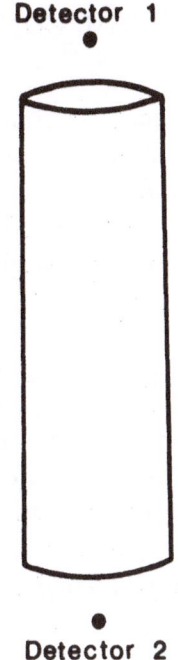

Detector 2

Figure 2.3.1 Two-detector configuration with a cylindrical sample.

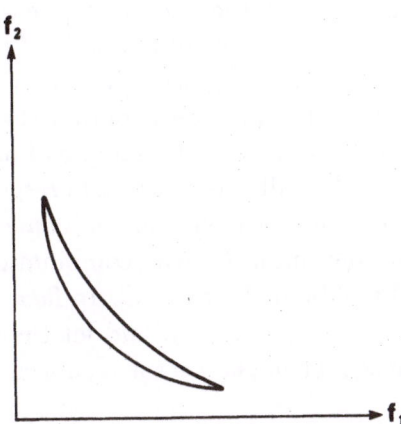

Figure 2.3.2 Schematic point-source response set.

2.4 THE CONVEXITY THEOREM: COMPLETE RESPONSE SETS

The point-source response set F contains all the values which the vector response function may attain in response to arbitrary location in the sample of a single point-source of unit intensity. Now let g and h be two elements of F, and let x and y be the source locations within the sample which give rise to these vector responses. That is

$$g = f(x) \qquad \text{and} \qquad h = f(y) \tag{2.4.1}$$

Now suppose a fraction $0 < a < 1$ of the unit source is located at position x and the remainder is located at position y. Then, by assuming linearity in mass, the vector response will be

$$f = ag + (1 - a)h \tag{2.4.2}$$

Unless F is a convex set, it is evident that f is not necessarily in F, as illustrated in figure 2.4.1 for the two-detector example discussed earlier. More generally, we see that the point-source response set does not necessarily contain the vector response to an arbitrary spatial distribution in the sample of one unit of source material.

However from these considerations we see that it is an easy matter to construct a set which does contain the vector response to any spatial distribution within the sample of one unit of source material. This set will be called the *complete response set*, denoted C. The complete response set will be the *convex hull* of F: the smallest convex set containing F. Any combination of elements of F such as in eq.(2) will be included in C. In fact, let a unit source be divided into quantities $a_1, a_2, \ldots$, and let the i-th quantity be positioned at point x^i, Then the vector response is

$$f = \sum a_i f(x^i) \tag{2.4.3}$$

It is evident that f is in the convex hull of F. Conversely any element of the convex hull of F can be written as in eq.(3) and thus

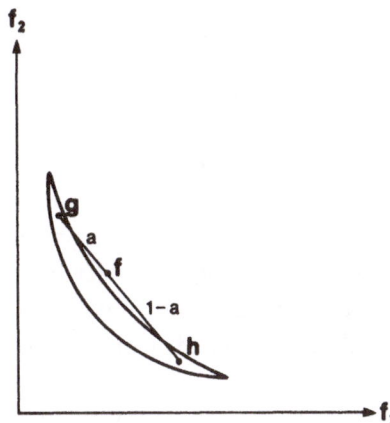

Figure 2.4.1 Schematic point-source response set.

corresponds to some spatial distribution of a unit source. The following Theorem summarizes the result which we have obtained.

THEOREM 1 The convex hull, C, of the point-source response set F contains the time-averaged vector response to any spatial distribution of one unit of source material. Furthermore, every element of C is the vector response to some spatial distribution of one unit of source material.

This simple result, known as the *convexity theorem*, is the foundation of the deterministic measure of performance to be developed in this Chapter.

It is an easy matter to extend the complete response set C to include any spatial distribution of an arbitrary quantity u of source material. We must assume or verify for the particular application in question that the assumption of linearity in mass holds for the quantity u. Then it is evident that the vector response for an arbitrary spatial distribution of u units of source material is just u times the vector response from a certain, analogous, spatial distribution of one unit of source material. Let uC be the set formed by multiplying each element of C by u. That is

$$uC = \{g : g = uc \quad \text{for all} \quad c \in C\} \qquad (2.4.4)$$

For any positive real u, the set uC contains the time-averaged vector response from any spatial distribution in the sample of u units of source material. Each element of uC corresponds to some spatial distribution of u units of source material. The matrix of the sample is assumed to be non-varying from sample to sample. In a similar fashion, by assuming linearity in time, we see that utC is comprised of the average number of counts obtained in duration t from any spatial distribution of u units of source material.

2.5 RELATIVE MASS RESOLUTION AND THE CONCEPT OF EXPANSION

In figure 2.5.1(a) we show the complete response set for one unit of source material derived from the point-source response set of figure 2.3.2. In figure 2.5.1(b) we superimpose on C the complete response set for u units of source material , where u is only slightly greater than unity. Because of the multiplicative relation between the two sets, one sees that uC is obtained by expanding C along the rays radiating from the origin and intersecting C. For the particular value of u chosen, C and uC overlap. In figure 2.5.1(c) is shown C and the complete response set vC for a greater quantity of source material, $v > u$. From this figure we see that v is sufficiently large so that C and vC do not overlap at all.

Each point in the set C corresponds to the vector response to some spatial distribution of one unit of source material. Likewise each element of uC represents the vector response to some spatial distribution of u units of source. Thus the overlapping response sets shown in figure 2.5.1(b) may be interpreted to mean that some spatial distributions of one unit of source material yield vector responses which are indistinguishable from the vector responses of some spatial distributions of u units of source material. In other words, 1 unit and u units of source material are not always distinguishable. We see that u is less than the relative mass resolution of the assay system whose complete response set is C. The situation

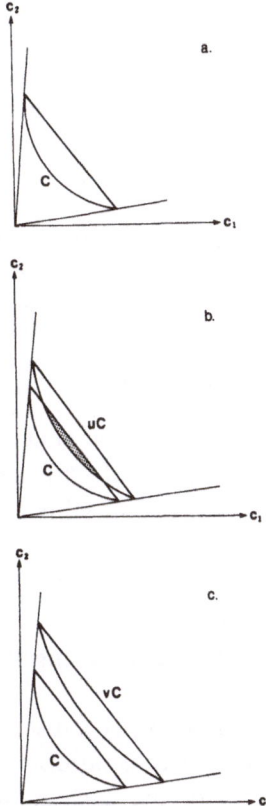

Figure 2.5.1 (a) The complete response set C derived from the point-source response set of figure 2.3.2. (b) C and the complete response set uC for a quantity u of source material, where u is only slightly greater than unity. (c) C and vC, where v is much greater than unity.

depicted in figure 2.5.1(c) indicates that every spatial distribution of 1 unit of source material yields a vector response distinguishable from every spatial distribution of v units of source material. In other words, v is greater than the relative mass resolution of the system.

These considerations suggest an immediate generalization. Let u and v be any quantities of source material. These quantities are *always distinguishable* if every spatial distribution of u units of source material yields a vector response distinct from every spatial distribution of v units. That is, u and v units are always distinguishable if and only if uC and vC do not overlap:

$$uC \cap vC = \emptyset \qquad\qquad (2.5.1)$$

where $\emptyset$ is the null set. Expressing this relation in terms of the individual elements of C, we state that the quantities u and v are always distinguishable if and only if there are no elements g and h in C such that

$$ug = vh \qquad\qquad (2.5.2)$$

This criterion of distinguishability is so general, and requires such tedious examination of C, as to be of little practical value. However, it suggests the following concept, which will prove to be quite useful.

It is easy to show that any two points g and h in the real n-dimensional Euclidean space $E(n)$ are related as in eq.(2) only if they lie on the same straight line radiating from the origin. This leads us to the following definition.

DEFINITION 1 Let G be a non-empty subset of the n-dimensional Euclidean space $E(n)$. For any real number y, the set yG is the set obtained from G by multiplying each element of G by y. The *expansion of G*, if it exists, is denoted $e(G)$ and is defined by

$$e(G) = \sup\{y : G \cap yG \neq \emptyset\}$$

That is, the expansion of G is the supremum of the set of real numbers y for which G and yG intersect. If this set is unbounded, then the expansion does not exist. This may occur in different ways. For instance, $e(G)$ does not exist is G contains the origin or if G contains any unbounded subset of a ray from the origin. On the other hand, unbounded sets may have finite expansion. For example, the set of points defining a hyperbola in $E(2)$ is unbounded but has expansion equal to unity.

If there are elements f and g in G such that

$$g = e(G)f$$

then G is said to be *self-expanded*. In the next Chapter we shall prove that if G is a closed and bounded set whose expansion exists, then G is self-expanded. The following Theorem, which will be proved in the next Chapter, shows our first application of the concept of expansion [8].

THEOREM 2 Let G be a closed and bounded convex set in $E(n)$ of finite expansion. Then for $m + d > m > 0$,

$$mG \cap (m + d)G = \emptyset$$

if and only if

$$\frac{m + d}{m} > e(G) \qquad .$$

The complete response set C is always convex and in practice will almost invariably be closed and bounded. Thus we may replace G in this Theorem by C. The Theorem now states that the quantities m and $m + d$ of source material are always distinguishable if and only if the ratio $(m + d)/m$ exceeds the expansion of the complete response set C. In other words the expansion of C is precisely equal to the relative mass resolution. The importance of this result is that we shall be able to develop an efficient computerizable algorithm for evaluating the expansion of C. To clarify these ideas we will now discuss three examples.

2.6 EXAMPLE: Pu ASSAY WITH ONE DETECTOR

We shall first examine a single-detector passive-gamma assay system. The sample is cylindrical, with height 60 cm and radius 10 cm. The matrix material is polyethylene with density 0.92 g/cm^3. The source material is ^{239}Pu, emitting gamma radiation at 384 keV. The linear absorption coefficient of the matrix is 0.10 cm^{-1}. We shall ignore the finite size of the detector, and adopt an idealized point-detector response function, given by

$$f(d, D) = \frac{E}{4\pi D^2} e^{-\mu d} \qquad\qquad (2.6.1)$$

where D is the total distance from the point source to the detector, d is the distance through the matrix of the sample between the source and the detector, E is the detection efficiency of the detector, and μ is the linear attenuation coefficient. Since we are considering a single detector the point-source response set F is one dimensional. Because of the simple contiguous structure of the sample, F comprises a single closed interval on the real line:

$$F = [f_{min}, f_{max}] \qquad\qquad (2.6.2)$$

Consequently F is a convex set, and thus precisely equals C, the complete response set. The expansion of F (and of C in this example) is given by the ratio

$$z = \frac{f_{max}}{f_{min}} \qquad\qquad (2.6.3)$$

which precisely equals the relative mass resolution of the assay system, as established by Theorem 2.

Since the sample is finite, f_{min} is also finite and non-zero. If the detector is positioned so as to touch the sample, then f_{max} diverges and the expansion is infinite. If the detector is removed from the sample then both f_{max} and z are finite.

Let us consider the values of f_{max}, f_{min} and z for varying detector positions along the quarter circular arc of radius 40 cm around the sample center, as shown in figure 2.6.1. The radial position and height of the detector are denoted R and Z, respectively. In figure 2.6.2 we show the values of f_{min} and f_{max}, which have been normalized to the value of f_{max} at $Z = 0$, $R = 40$ cm. The point-source position in the sample which yields the maximum response moves along the sample wall facing the detector, as denoted by the open circle in figure 2.6.1. The source position yielding the minimum response is at the bottom of the sample on

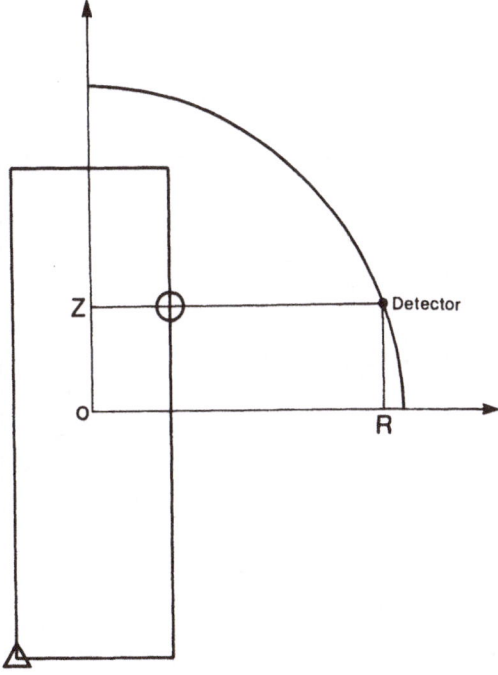

Figure 2.6.1 Locus of detector positions on a quarter-circular arc around the cylindrical sample.

the side furthest from the detector, as indicated by the open triangle in figure 2.6.1. We note that f_{max} increases, throughout the range of positions shown, as the detector rises from the sample midplane. This is because the detector continuously approaches the sample wall. The minimum value of the response decreases because both the source-detector distance and the matrix thickness increase.

Also shown in figure 2.6.2 is the relative mass resolution z as expressed by eq.(3). When the detector is on the midplane, the relative mass resolution equals 38.9. That is, every spatial distribution of u grams of source material will be distinguishable from every spatial distribution of v grams if and only if

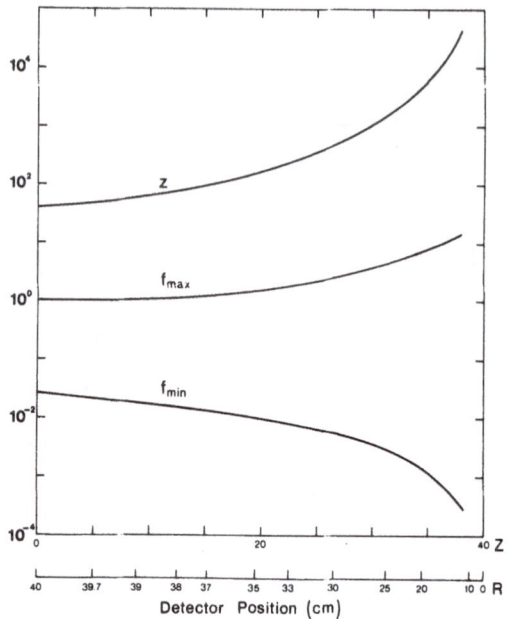

Figure 2.6.2 Normalized minimum and maximum detector response and relative mass resolution versus detector position.

$$v > 38.9u \qquad\qquad (2.6.4)$$

We see from the figure that the resolution steadily deteriorates as the detector rises, as expressed by the rising value of the expansion. The reason for this is that the rate of decrease of f_{min} exceeds the rate of increase of f_{max}, as the detector position rises.

The relative mass resolution of 38.9 obtained when the detector is on the sample midplane is indeed quite poor, and can be improved by placing the detector further from the sample. In figure 2.6.3 we show f_{max} and f_{min} versus the radial position of the detector on the midplane. f_{max} decreases more rapidly with R than f_{min}, causing the resolution to improve with increasing radial position. At 100 cm from the sample center the expansion has decreased to 12.8. The asymptotic value of f_{max} is significantly greater than the limiting value of f_{min} in our example, due

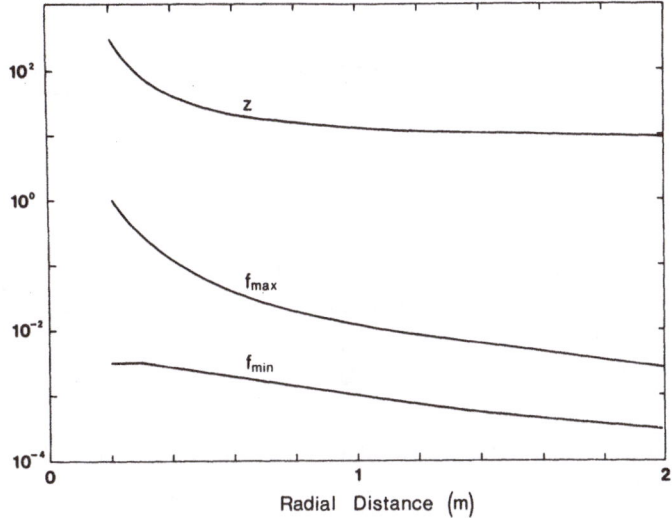

Figure 2.6.3 Normalized minimum and maximum detector response and relative mass resolution versus radial position of the detector on the sample midplane.

to the strong absorption of the detected radiation in the polyethylene matrix of the sample. Examination of eq.(1) shows that at infinite sample-detector separation the expansion approaches the asymptotic value

$$z_{\text{asym}} = e^{2\mu r_0} \qquad (2.6.5)$$

where r_0 is the sample radius. The asymptotic value of the expansion is 7.4. Thus at infinite separation every spatial distribution of u grams of source material is distinguishable from every spatial distribution of v grams if and only if

$$v > 7.4u \qquad (2.6.6)$$

We have not yet considered the very important factor of statistical uncertainty. Indeed, when we consider removing the detector to a great distance from the sample we must necessarily consider the resulting reduction in the count rate and the associated increase in statistical uncertainty. We shall treat this in section 2.11.

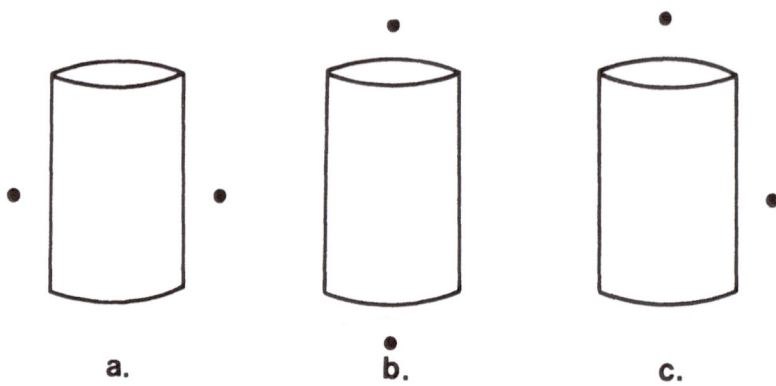

Figure 2.7.1 Three different two-detector deployments around a cylindrical sample.

Before doing so, let us extend the present example to include an additional detector.

2.7 EXAMPLE: Pu ASSAY WITH TWO DETECTORS

We shall now employ two detectors whose response functions are given by eq.(2.6.1). An infinite variety of deployments of these two detectors around the sample is conceivable. We shall consider the three characteristic arrangements shown in figure 2.7.1. In every case each detector is 40 cm from the center of the sample. In configuration (a) both detectors are on the sample midplane and separated by 180 degrees. In configuration (b) both detectors are on the rotational symmetry axis of the sample, one above and one below the sample. In configuration (c) one detector is on the rotational symmetry axis and one is on the sample midplane.

We may evaluate the relative mass resolution of each configuration by plotting the point-source response set F, using F to construct the complete response set C, and manually evaluating the expansion of C. In later sections of this Chapter we will develop a much more elegant and computationally convenient method. However the graphical approach is adequate for a two-detector config-

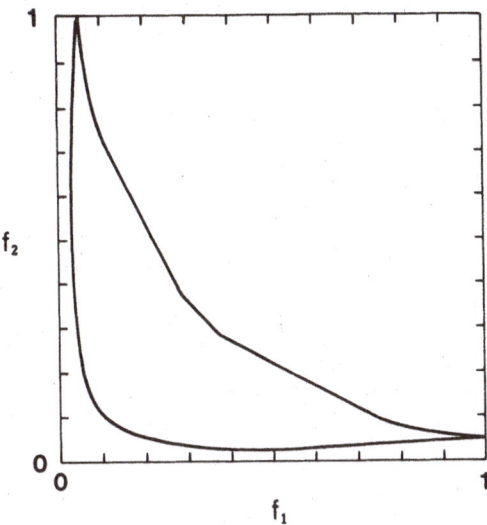

Figure 2.7.2 Point-source response set for the two-detector configuration (a) shown in figure 2.7.1. The responses have been normalized to their maximum values.

uration, and will illustrate the basic features of the more general procedure.

To construct the response set, we must evaluate the response function for each detector for a range of point-source positions in the sample. Let f_1 and f_2 represent the response functions for the first and second detector. For each point-source position in the sample we plot the point (f_1, f_2) on the f_1, f_2 plane. The locus of these points defines the point-source response set [9].

Figure 2.7.2 shows the boundary of the response set for detector configuration (a). We see that F is not a convex set, though its convex hull, C, is constructed by adding a single straight line, as shown in figure 2.7.3. We employ the complete response set C to evaluate the expansion, or relative mass resolution. We recall that the expansion is the largest number z such that there are elements g and h in C satisfying

$$h = zg \qquad (2.7.1)$$

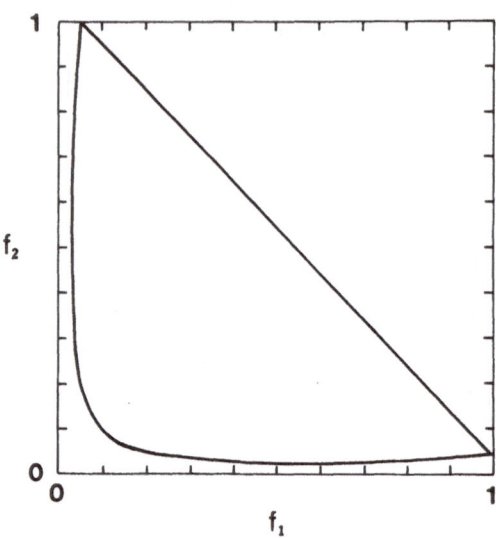

Figure 2.7.3 Complete response set for the two-detector configuration (a) shown in figure 2.7.1. The responses have been normalized to their maximum values.

In order for two points on the plane to satisfy a relation such as this they must lie on the same ray radiating from the origin. There is some ray whose points of intersection with the boundary of C are related, as in eq.(1), by a factor z which is greater than (or at least not less than) for any other ray. This factor is the expansion of C. By trial and error we find the expansion of the complete response set in figure 2.7.3 to be 5.1, occurring on the ray which bisects C [10]. Thus for two detectors located on the sample midplane on opposite sides of the sample at 40 cm from the sample center the relative mass resolution is 5.1. In other words, any spatial distribution of u grams can be distinguished from any spatial distribution of v grams if and only if

$$v > 5.1u \qquad (2.7.2)$$

This is much better resolution than the value of 38.9 obtained for a single detector at 40 cm on the midplane. In fact two detectors at 40 is better than one detector at infinity, for which the relative mass resolution is 7.4.

A similar procedure may be pursued for detector configurations (b) and (c). For two detectors on the rotational symmetry axis the relative mass resolution is 187. This is much better than for a single detector on the symmetry axis (off scale in figure 2.6.2), though considerably poorer than two midplane detectors. With one detector on the midplane and one on the symmetry axis the relative mass resolution is 38.9 — indistinguishable from the case of a single midplane detector. This is not surprising: the axial detector is of negligible utility in comparison to the midplane detector.

Our analysis so far indicates that the best arrangement is configuration (a). However a relative mass resolution of 5.1 may be far from adequate. We can improve the two-detector resolution by removing the detectors to a greater distance. As in the one-detector case the relative mass resolution converges to a value greater than unity at infinite sample-detector separation [11]. In this example the asymptotic value of the expansion is 2.72. When considering large sample-detector distances it is essential to consider the statistical uncertainty, which we shall do in section 2.11.

Since dramatic improvement in resolution was obtained by using two detectors rather than one, we might expect significant improvement with three or more detectors. When we contemplate the analysis of an assay system with more than two detectors, it soon becomes evident that the graphical method which we employed in this section is inadequate. It is arduous to construct and visualize the multi-dimensional response set, and even more difficult to manually search for the expansion of its convex hull. We have by now acquired some feeling for the concept of expansion — its graphical or geometrical manifestation and its physical significance. In section 2.9 we shall turn to the problem of developing an efficient computerizable algorithm for evaluating the expansion. Before relegating the computation of the relative mass resolution to the computer, we shall consider one more example for which the graphical method is illuminating.

2.8 EXAMPLE: COINCIDENCE MEASUREMENTS

In this section we shall apply the convexity theorem to a distinctive type of assay problem: the measurement of fissile material by detecting coincident radiations. A unique feature of the fission process is the large number of long-range particles emitted at the time of fission: typically about eight photons and neutrons [12]. This multiplicity of emitted particles has been exploited with great success in the assay of fissile materials [13, 14]. Measuring the rate of coincident events between different detectors yields two basic advantages. First, by eliminating uncorrelated background events one achieves considerable improvement in the peak-to-background ratio. Second, simultaneous use of coincidence measurements of different orders appreciably decreases the sensitivity of the assay to the random spatial distribution in the sample of the fissile material.

Spatial randomness of the fissile material in the matrix of the sample is often the greatest source of uncertainty in the assay. This uncertainty is progressively reduced by increasing the number of detectors and by using the full range of available coincidence measurements. Unfortunately, the cost and complexity of the assay increase with each additional detector. The basic task in the design of such an assay system is determination of the number of detectors to use, and what orders of coincidence-multiplicity to examine.

The traditional theoretical approach to assay-system design is based on evaluating the efficiency of the multiple-coincidence measurement for selected spatial distributions of the fissile material. Mathematical expressions involving integrals on the spatial distributions of the source material in the sample are derived. However, these integrals are extremely difficult to evaluate, even for two-fold coincidence, and meaningful results are usually obtained by making rather strong assumptions about the nature of the assay problem [15]. Furthermore, a fundamental difficulty with this approach is that one has no way of evaluating the required integrals for all of the uncountably infinite number of possible spatial distributions.

The attempt to identify a limited number of "characteristic" spatial distributions is problematic since there is no clear or certain way of knowing whether or not accurate evaluation is made of the sensitivity of the measurement to the full range of possible spatial distributions of the fissile material.

These limitations of the classical theoretical approach can be avoided by experimental investigation of various alternative assay-system designs. Indeed this approach has yielded very useful and meaningful results [16]. However, experimental work is time consuming and expensive. Applying the convexity theorem and the concept of relative mass resolution circumvents the limitations of the classical approach to design, and yet requires limited experimental data. This design-analysis yields precise evaluation of the relative mass resolution of the assay system for an arbitrary number of detectors, and rigorously accounts for all possible spatial distributions of the fissile material.

In this section we shall apply this technique to the analysis of systems for the assay of low-density spatially-random fissile materials by means of coincidence measurements. In section 2.8.1 we present the precise formulation of the assay problem. In sections 2.8.2 and 2.8.3 we study the one- and two-detector systems, and in section 2.8.4 we examine the n-detector system in which we use only the $(n-1)$-fold and n-fold coincidence measurements. Our results will be completely analytical expressions for the relative mass resolution.

2.8.1 Formulation of the Assay Problem

Let the vector x represent an arbitrary point within the sample, and let X represent the set of all points in the sample. Thus for instance, if the sample is a right cylinder of height $2H$ and radius R and centered at the coordinate origin, then points in the sample can be conveniently represented with cylindrical coordinates, and

$$X = \{(r, \theta, z) \quad \text{for all} \quad 0 \le r \le R,\ 0 \le \theta \le 2\pi,\ -H \le z \le H\}$$

$$(2.8.1)$$

The sample contains a non-fissile *matrix* material whose spatial distribution within the sample does not vary from sample to sample. The simplest case of a non-varying matrix is the homogenous matrix. In addition to the matrix material the sample contains an unknown amount of fissile *source* material, whose assay we seek. The spatial distribution of the fissile material in the sample is random. That is, any spatial distribution is possible, though we need not specify the likelihood of occurrence of any given spatial distribution. The matrix material emits no relevant radiations and yet absorbs the radiations spontaneously emitted by the fissile source material. The fissile material emits on the average N particles in each fission. For simplicity we shall assume that all fissions emit precisely N particles. In practice these particles need not all be of the same type: typically both photons and neutrons of various energies reach the detector. We shall however ignore this diversity and assume that all the emitted particles are identical. Elimination of these assumptions presents no fundamental difficulty; we shall adhere to them for the sake of clarity and simplicity. In contrast to this, the problem becomes much more difficult if the spatial distribution of matrix material (as well as of the fissile material) may vary from sample to sample. This situation may be handled by the techniques developed in Chapter 3.

An array of n detectors are positioned around the sample. These detectors need not have similar characteristics, nor need they be symmetrically placed with respect to the sample. Each detector is fully characterized by its response function, $f_i(x)$, which expresses the time-averaged count rate of the detector in response to a point-source of unit intensity at position x in the sample. The response is typically a smooth and continuous function of the point-source position, as shown in figure 2.8.1. The concentration of the fissile material in the sample is sufficiently low so that the absorption of radiation in traversing any arbitrary segment of the sample is unaffected by the spatial distribution of the fissile material. However, absorption in the matrix is fully accounted for by the response function. Furthermore, the response in each detector

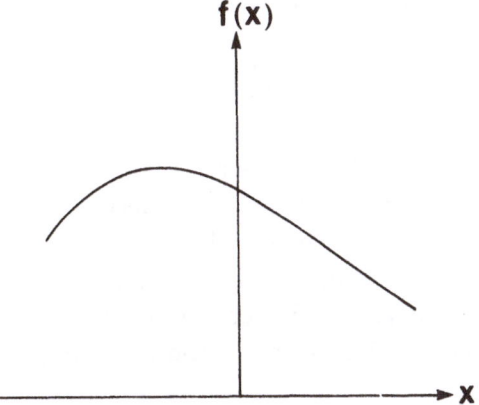

f(x)

Figure 2.8.1 Schematic response function (counts/g sec) for a detector parallel to a line source of length $2H$. Midplane of detector and sample is at $x=0$.

to any spatial distribution of source material is a linear combination of the responses to each local deposit. That is, if the density of source material at point x in the sample is $r(x)\,\mathrm{g/cm^3}$, then the average response in detector i is

$$R_i = \int f_i(x)r(x)\,dx \qquad \text{counts/sec} \qquad (2.8.2)$$

Now let us suppose that a fission event has occurred at point x in the sample. We shall assume that the probability is negligible that more than one particle from this fission event will reach a given detector. Let A be the specific activity of the source material. Then the ratio $f_i(x)/A$ is the probability that the i-th detector will sense some particle emitted in the fission event. The probability that the i-th detector will sense a particular one of these particles is

$$p_i(x) = \frac{f_i(x)}{NA} \qquad (2.8.3)$$

Since each fission event leads to the essentially simultaneous emission of N particles, we will observe a certain rate of simultaneous detections in any subset of $m \leq N$ detectors. Let

$$I(m) = \{i_1, i_2, \ldots, i_m\} \qquad (2.8.4)$$

represent the set of subscripts of the m detectors whose simultaneous detection we wish to observe. For instance we may wish to follow the rate of two-fold coincidence between detectors 3 and 4. Then $m = 2$ and $i_1 = 3$ and $i_2 = 4$.

Given that a fission event has occurred at point x, the probability that detector i_1 will detect one of the N emitted particles, that detector i_2 will detect one of the remaining $N - 1$ particles, etc., is

$$p_{I(m)}(x) = N(N-1)\cdots(N+1-m)p_{i_1}(x)p_{i_2}(x)\cdots p_{i_m}(x) \qquad (2.8.5)$$

Now the time-averaged coincident count rate in these m detectors in response to a unit point source at position x in the sample is given by the m-fold coincidence response function as

$$f_{I(m)}(x) = A p_{I(m)}(x) \qquad (2.8.6)$$

$$= A \frac{N!}{(N-m)!} \prod_{k=1}^{m} p_{i_k}(x) \qquad (2.8.7)$$

From eq.(6) we see that the m-fold coincidence rate is proportional to the specific activity of the source deposit, and thus will display linear superposition just as the singles rate does, as expressed in eq.(2). Employing eq.(3) this relation becomes

$$f_{I(m)}(x) = \frac{(N-1)!}{(N-m)!\,(NA)^{m-1}} \prod_{k=1}^{m} f_{i_k}(x) \qquad (2.8.8)$$

In many applications, the symmetry of the sample-detector arrangement makes all the detectors equivalent, and thus causes all the single-detector response functions to be equal. The common response function will be denoted $f(x)$. In this high-symmetry case we have no reason to single out any particular m-fold coincidence, and are usually interested only in the rate at which *some* m-fold coincidence occurs among the n detectors. This rate is given by

$$h_m(x) = C(n,m) f_{I(m)}(x) \qquad (2.8.9)$$

$$= r_m f^m(x) \qquad (2.8.10)$$

where $C(n,m)$ is the binomial coefficient and

$$r_m = \frac{(N-1)!\, C(n,m)}{(N-m)!\,(NA)^{m-1}} \qquad (2.8.11)$$

The information obtained from an assay system employing n symmetric detectors is contained in the m response functions $h_1, h_2, \ldots, h_m$. Experience in this field has indicated that considerable improvement in resolution is obtained by employing coincidence measurements. This arises from the fact that the coincidence measurements are very efficient at simplifying the complexities which arise from spatial randomness of the source material. However, the expense and difficulty of maintaining and operating the assay system and the complexity of interpreting the measurements all increase with the number of detectors employed. In the remainder of this example we shall illustrate the procedure for evaluating the mass resolution capability of the assay system as a function of the number or arrangement of the detectors. This provides absolute quantitative information on the capability of each assay system, as well as evaluation of the utility of the marginal detector. The sample geometry which we shall examine will be sufficiently simple so that we shall obtain entirely analytical results. This will lead to further understanding of how the convexity theorem is applied, and will illuminate the mechanism by which coincidence information "sorts out" the complications imposed by spatial randomness of the source material.

The sample-detector geometry which we shall examine is the following: The sample is a straight rod with fissile material distributed in a homogeneous matrix. The rod may be enclosed in a shield material. This makes no difference for our analysis provided the absorption in the shield is accounted for in the response function. The rod is sufficiently thin so that the spatial distribution of the source material is essentially one-dimensional along the length of the rod. The fissile material may for example be ^{235}U or ^{239}Pu, while the matrix material may be ^{238}U. In any case the radiation-absorption properties of the source and of the matrix material are very similar. Consequently, the absorption of radiation in traversing any segment of the sample is unaffected by the spatial distribution of the fissile material. Thus the assumption of linear superposition of responses — eq.(2) — is valid. The detectors are placed symmetrically around the sample on its midplane. This assay problem is not as hypothetical as it may seem, and has been subjected to extensive experimental study.

We shall begin by examining the relative mass resolution of the one-detector and two-detector assay systems. From this we will see the remarkable improvement obtained by using the coincidence rate between the two detectors. Then we shall examine the utility of using the $(n - 1)$-fold and n-fold coincidences.

2.8.2 One-Detector Assay Systems

The response function of each detector, $f(x)$, may look something like that in figure 2.8.1. For our purposes the important point is that $f(x)$ — the time-averaged response to a unit point-source located at any position x in the sample — may assume any value between lower and upper limits, a and b respectively. By knowing the response $f(x)$ to a single point source as a function of position, the convexity theorem specifies the range of responses to *any* spatial distribution of the same total quantity of source material. Because of the linear superposition of the response, it is evident that the time-averaged response to any spatial distribution of one unit of source material will fall in the interval $[a, b]$.

Likewise, the time-averaged response to any spatial distribution of u grams of source material will fall in the interval $[ua, ub]$.

Now, consider two quantities of source material, u and v grams, where $v > u$. In order for any spatial distribution of v grams to yield a measurement distinguishable from every spatial distribution of u grams, we require that the smallest possible measurement obtainable from v grams exceed the greatest possible measurement of u grams. That is

$$va > ub \qquad (2.8.12)$$

The relative mass resolution of the one-detector system is defined to be the smallest number z_1 such that, for arbitrary mass u, any spatial distribution of $v > z_1 u$ grams is distinguishable from any spatial distribution of u grams. From eq.(12) we see that the relative mass resolution of the one-detector system is

$$z_1 = \frac{b}{a} \qquad (2.8.13)$$

2.8.3 Two-Detector Assay Systems

The two-detector assay system yields three measurements: the "singles" rates in each detector, f_1 and f_2, and the coincidence or "doubles" rate h_2. In this section we shall investigate the amount of additional information obtained by employing the coincidence measurement in the assay of random spatial distributions of fissile material in a long rod-shaped sample.

From eqs.(10) and (11) we note that the ratio of the product of the singles rates to the doubles rate, in response to a single point source, is independent of the location of the source:

$$\frac{f_1 f_2}{h_2} = \frac{NA}{N-1} \qquad (2.8.14)$$

As pointed out by Gozani [13], this relation is a strong indication that utilization of the coincidence information, in conjunc-

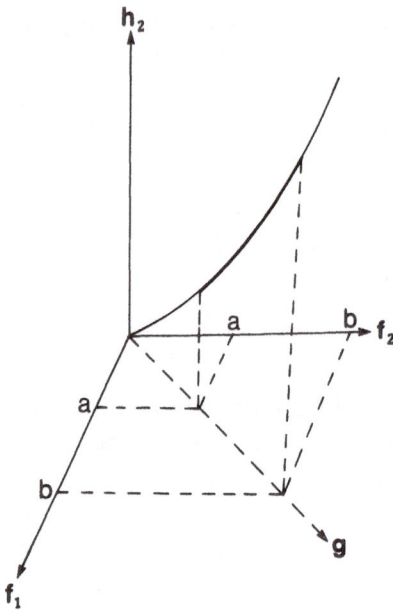

Figure 2.8.2 The point-source response set F for a two-detector assay system.

tion with the singles rates, will yield considerable improvement in mass resolution over the one-detector system. In other words, the coincidence measurement, together with the singles data, should be much less sensitive to random spatial distribution of the source material than is a single measurement. Indeed if the source material could appear only as a single point-deposit, eq.(14) shows that the singles and doubles data, when taken together, are entirely independent of the position of the point source, and thus would supply a completely precise assay (ignoring, of course, considerations of counting statistics). However, our assay problem is more complicated, since arbitrary spatial distribution of the source material is allowed. We shall see that the two-detector system is appreciably better than the one-detector case, though it is not perfect.

Let us now envision a three-dimensional response-space whose coordinate axes are f_1, f_2 and h_2. Each point x in the sample determines a point $(f_1(x),\ f_2(x),\ h_2(x))$ in this space, and the set of all such points is the point-source response set, denoted F. Every possible triplet of measurements in response to a single point-source is represented as an element in the set F. We schematically show the set F by the heavy parabolic segment in figure 2.8.2. Since the two detectors are equivalent, their singles rates are equal, so

$$f_1(x) = f_2(x) = f(x) \qquad (2.8.15)$$

From eq.(14) we see that

$$h_2 = \frac{N-1}{NA}\, f(x)^2 \qquad (2.8.16)$$

Thus the point-source response set is defined by a parabolic curve, whose projection on the $f_1 f_2$ plane is a line segment on the ray g bisecting the angle between the f_1 and f_2 axes. In fact F is entirely contained in the plane defined by the h_2 axis and the ray g. In figure 2.8.3 we plot F on this plane, where F is the locus of points defined by

$$h_2 = \frac{N-1}{2NA}\, g^2 \quad , \qquad g_1 \leq g \leq g_2 \qquad (2.8.17)$$

$$g_1 = \sqrt{2}\, a \quad , \qquad g_2 = \sqrt{2}\, b \qquad (2.8.18)$$

The response to a single unit point-source at any location in the sample is represented by a point in F. Referring to figure 2.8.4, let the point P represent the response to a unit point-source located at position x in the sample, and let the point Q represent the response to a unit point-source at position y. Now suppose the unit source is divided into two equal portions, one portion located at point x and the other at point y. Because of the linear superposition of the responses, the net response will be precisely at the midpoint, C, of the line joining P and Q. By performing unequal subdivisions of the unit source and positioning one portion

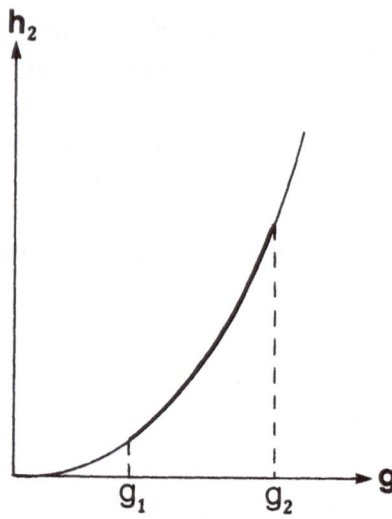

Figure 2.8.3   The point-source response set F for a two-detector system, in the h_2–versus–g plane.

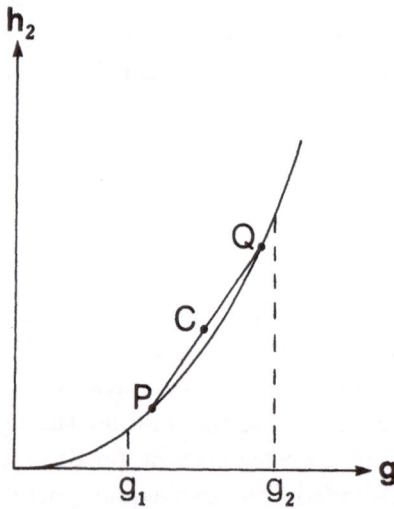

Figure 2.8.4   Geometrical construction showing the response to two point-sources each of 1/2 unit intensity.

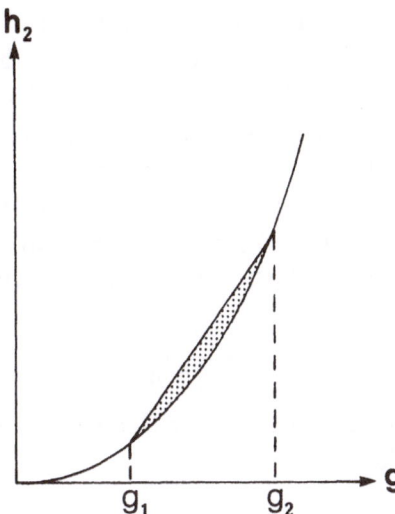

Figure 2.8.5 The shaded convex region represents the complete response set: all the responses obtainable from any spatial distribution in the sample of one unit of source material.

at x and the other at y, we obtain responses represented by every point on the line segment PQ. By applying this procedure to other points in the sample we realize that the response, not only to arbitrary binary divisions of a point source, but to any possible spatial distribution of one unit of source material, is represented by a point in the shaded convex region shown in figure 2.8.5. This convex region is the complete response set. In our example the complete response set is the region bounded by the parabola of eq.(17) and the straight line L defined by

$$h_2 = sg + t \qquad (2.8.19)$$

where

$$s = \frac{N-1}{2NA}(g_1 + g_2) \qquad (2.8.20)$$

and

$$t = -\frac{N-1}{2NA}g_1 g_2 \qquad (2.8.21)$$

This assay system is based on measuring three quantities: the rate of "singles" events in each detector and the rate of coincidences between the two detectors. The relative mass resolution of this two-detector system is the smallest number $z_{1,2}$ such that, for arbitrary mass u, every spatial distribution of $v > z_{1,2}u$ grams yields a triplet of measurements different from the triplet of measurements obtained from every spatial distribution of u grams. It is evident that $z_{1,2}$ is the largest value of z for which the parabola

$$h_2 = z \frac{N-1}{2NA} g^2 \qquad (2.8.22)$$

just touches the line L. In order for this parabola to intersect the line L, z must satisfy

$$z \frac{N-1}{2NA} g^2 = sg + t \qquad (2.8.23)$$

Thus the relative mass resolution may be expressed as

$$z_{1,2} = \underset{g_1 \le g \le g_2}{\text{maximum}} \frac{sg + t}{\frac{N-1}{2NA} g^2} \qquad (2.8.24)$$

An explicit expression for this maximum may be found, and is

$$z_{1,2} = \left(\frac{a+b}{2\sqrt{ab}} \right)^2 \qquad (2.8.25)$$

In other words, $z_{1,2}$ is the square of the ratio of the arithmetic mean to the geometric mean of a and b. From the arithmetic mean - geometric mean inequality we conclude that $z_{1,2}$ is greater than unity unless $a = b$. Furthermore one can readily see that $z_{1,2}$ is less that z_1. Summarizing, we conclude that

$$1 \le z_{1,2} \le z_1 \qquad (2.8.26)$$

with equalities if and only if $a = b$. Thus the coincidence measurement improves the mass resolution, but does not completely overcome the ambiguity imposed by the spatial randomness of the source material. For example, if the relative mass resolution with one detector is 1.3, then the relative mass resolution using the co-incidence of two detectors may be found with the aid of eqs.(13) and (25) to be 1.017. In other words, if with one detector, the ambiguity in determining the source mass is 30% due to spatial randomness of the source material, then it is only 1.7% when exploiting singles and coincidence data with two detectors.

2.8.4 Multi-Detector Systems: High-Order Coincidences

Consider an assay system with $n > 1$ symmetrical detectors. For instrumental simplicity we may wish to use only the $(n-1)$- and n-fold coincidence data. Additionally, when $n > 2$, we may wish to avoid using the singles data because of the large statistical uncertainty associated with them. When employing only the two high-order coincidence measurements, the point-source response set F is the set of all pairs of $(n-1)$- and n-fold measurements obtainable from any positioning in the sample of a unit point-source. Referring to eqs.(10) and (11), we see that F is the set of ordered pairs

$$F = \left\{ (r_{n-1} f^{n-1}, r_n f^n) \quad \text{for all} \quad a \le f \le b \right\} \qquad (2.8.27)$$

where a and b are respectively the minimum and maximum single-detector responses to a unit point source. Thus just as for the two-detector case in figure 2.8.3, the point-source response set can be represented as part of a polynomial curve in the plane. However, as n increases, the curvature decreases: F approaches a straight line segment. This means that the convex region (as in figure 2.8.5) which represents the pairs of responses from all possible spatial distributions of a unit source, becomes a thin slice as n increases. We should thus expect that the relative mass resolution, $z_{n-1,n}$, will decrease (that is, improve) as n rises.

Employing a line of reasoning similar to that for the two-detector case, one can develop an expression for $z_{n-1,n}$ in terms

Table 2.8.1

Relative Mass Resolution
Using $(n-1)$-fold and n-fold Coincidences

z_1	$z_{1,2}$	$z_{2,3}$	$z_{3,4}$	$z_{4,5}$	$z_{5,6}$	$z_{6,7}$	$z_{7,8}$
1.2	1.0083	1.0031	1.0018	1.0013	1.0010	1.0008	1.0007
1.5	1.042	1.015	1.0091	1.0064	1.0049	1.0040	1.0034
2.0	1.125	1.046	1.027	1.019	1.014	1.011	1.0098
5.0	1.800	1.257	1.146	1.102	1.077	1.062	1.052
10.0	3.025	1.559	1.307	1.209	1.158	1.126	1.105

of a and b, or more conveniently, in terms of the relative mass resolution for a single detector, z_1. The final result is

$$z_{n-1,n} = \frac{n-1}{n^q} \frac{z_1^q - 1}{z_1 - 1} \left(\frac{z_1^q - 1}{z_1^q - z_1} \right)^{q-1} \tag{2.8.28}$$

where

$$q = \frac{n}{n-1} \qquad n = 2, 3, 4, \ldots \tag{2.8.29}$$

In Table 2.8.1 we present the relative mass resolution obtained with $(n-1)$- and n-fold coincidence, versus n, for various values of the relative mass resolution with 1 detector. The immediate conclusion obtained from this table is that $z_{n-1,n}$ converges to unity as n rises: rapidly at first, gradually for larger n. Indeed it is impressive that when the relative mass resolution with one detector is 2.0 (that is, 100% uncertainty), with two detectors the relative mass resolution is 1.125 or 12.5% uncertainty, and with three detectors the relative mass resolution is 1.046 or 4.6% uncertainty. The marginal utility of the fourth detector is very small, as expressed by the fact that the relative mass resolution with four detectors is 1.027 — only slightly less than with three detectors.

We have obtained explicit analytical expressions for the relative mass resolution by exploiting the high degree of symmetry and the simple geometry of our example. However, if we wished to find an explicit expression for the relative mass resolution obtained when using all the coincidence measurements rather than just the two coincidence rates of highest order, we would confront a rather complex problem. Likewise if we wish to analyze the resolution capability for sample shapes other than the long thin rod, we would find the mathematical burden to be considerable. In order to handle these more complex situations a general computerizable algorithm has been developed for numerical evaluation of the relative mass resolution. This algorithm is the subject of the next section.

2.9 COMPUTATION OF THE RELATIVE MASS RESOLUTION

In this section we shall present the details of an efficient computerizable algorithm for evaluating the relative mass resolution for an arbitrary number of detectors [17]. The algorithm depends on evaluating the expansion of the complete response set C. We begin by demonstrating (intuitively) how one evaluates the expansion of C if the elements of C are explicitly known. C is the convex hull of the point-source response set F, and usually only F is explicitly known. Thus it is necessary to be able to evaluate the expansion of C directly from F. We shall discuss this briefly before summarizing the algorithm for evaluating the relative mass resolution.

We will use the succinct notation $\langle a , b \rangle$ for the inner product of two n-tuples $a_1, \ldots, a_n$ and $b_1, \ldots, b_n$. That is,

$$\langle a , b \rangle = \sum_{i=1}^{n} a_i b_i$$

2.9.1 Intuitive Derivation

The complete response set C is a closed and bounded convex set in the n-dimensional real Euclidean space $E(n)$. A plane in $E(n)$ may be defined as the locus of points $x_1, \ldots, x_n$ which satisfy $\langle a, x \rangle = b$ for constants $a_1, \ldots, a_n$ and b. The quantity b may be varied to produce a family of parallel planes.

Let $a_1, \ldots, a_n$ be real numbers defining parallel planes in $E(n)$ such that for every element c in C

$$\langle a, c \rangle \neq 0 \qquad (2.9.1)$$

In order for there to be any n-tuples satisfying relation (1), it is necessary that C not contain the origin. Any point c in C belongs to a particular plane in the family, whose distance from the origin is given by

$$d(c) = \frac{|\langle a, c \rangle|}{\sqrt{\langle a, a \rangle}} \qquad (2.9.2)$$

Thus condition (1) states that the n-tuple $a_1, \ldots, a_n$ defines planes which do not intersect both C and the origin. The plane defined by any point c is denoted $P(c)$. Since C is a closed and bounded set, there is an element u in C such that no plane defined by a point in C is closer to the origin than $P(u)$. Likewise, there is a point v in C such that no plane defined by a point in C is further from the origin than $P(v)$. We may algebraically express this property of u and v by stating:

$$|\langle a, u \rangle| = \underset{c \in C}{\text{minimum}} |\langle a, c \rangle| \qquad (2.9.3)$$

$$|\langle a, v \rangle| = \underset{c \in C}{\text{maximum}} |\langle a, c \rangle| \qquad (2.9.4)$$

It is evident that any ray radiating from the origin which intersects $P(u)$ also intersects $P(v)$. Furthermore, every point p on $P(u)$, when multiplied by the ratio $|\langle a,v\rangle/\langle a,u\rangle|$ yields the point on $P(v)$ which is contained in the ray containing p. In other words, the set of points $P(v)$ is an element-by-element multiple of the set of points $P(u)$, according to the relation

$$P(v) = \frac{|\langle a,v\rangle|}{|\langle a,u\rangle|}\, P(u) \tag{2.9.5}$$

Now, C is a convex set and the points u and v of eqs.(3) and (4) define extremal supporting planes. In other words, C is entirely "above" the plane $P(u)$ and entirely "below" the plane $P(v)$. Upon consideration of eq.(5), it is evident that for any number $b > 1$

$$C \cap bC = \emptyset \tag{2.9.6}$$

if and only if

$$b > \frac{|\langle a,v\rangle|}{|\langle a,u\rangle|} \tag{2.9.7}$$

From Theorem 2, the expansion of C is the smallest number z such that

$$C \cap yC = \emptyset$$

for all $y > z$. Hence the expansion of C is the smallest value of the ratio in ineq.(7) which may be obtained from any n-tuple a_1, ..., a_n satisfing relation (1). That is, the expansion of C is given by

$$e(C) = \underset{a_1,\ldots,a_n}{\text{minimum}}\ \frac{|\langle a,v\rangle|}{|\langle a,u\rangle|} \tag{2.9.8}$$

where u and v are elements of C satisfying eqs.(3) and (4), and the n-tuples a_1, ..., a_n on which the minimization is performed satisfy relation (1).

Eq.(8) suggests an iterative algorithm for calculating the expansion of the response set C. We begin by choosing an initial n-tuple, a_1, ..., a_n. Then we search for the elements u and v of

C which, respectively, minimize and maximize the absolute value of the weighted sums as in eqs.(3) and (4). Verification that the n-tuple satisfies relation (1) is easily done, as we shall show in Lemma 1. Then we systematically alter the n-tuple, seeking new extremal points u and v for each n-tuple, until the minimum ratio in eq.(8) is obtained. This ratio is precisely equal to the expansion of C.

The only difficulty with this algorithm is that it requires a search for a minimum on the complete response set C, which is the convex hull of the point-source response set F. In practice we can readily calculate F, but if we are dealing with more than two dimensions it is usually not easy to explicitly construct the convex hull of F.

This difficulty is easily overcome. Since C is the convex hull of F, any plane which is an extremal plane of C will also be an extremal plane of F and vice versa. Thus eqs.(3), (4) and (8) when applied to F (rather than to C) will also yield precisely the expansion of C. These relations may be succinctly combined as

$$e(C) = \operatorname*{minimum}_{a \in A} \; \operatorname*{maximum}_{f,g \in F} \frac{|\langle a,f \rangle|}{|\langle a,g \rangle|} \qquad (2.9.9)$$

where A is the set of all real n-tuples for which

$$\langle a,f \rangle \neq 0 \qquad \text{for all} \qquad f \in F \quad . \qquad (2.9.10)$$

The following Lemma shows how one may verify that a given n-tuple $a_1, \ldots, a_n$ in fact satisfies relation (1).

LEMMA 1   Let G be a convex set in $E(n)$. Then the following two statements are equivalent.

$$\langle a,g \rangle \neq 0 \qquad \text{for all} \qquad g \in G \qquad (i)$$

$$\langle a,f \rangle \langle a,g \rangle > 0 \qquad \text{for all} \qquad f \text{ and } g \in G \quad . \qquad (ii)$$

PROOF   Statement (ii) clearly implies statement (i). To prove the converse, suppose that there are elements b and c in G such that

$$\langle a,b \rangle > 0 \qquad \text{and} \qquad \langle a,c \rangle < 0 \quad .$$

There is a real number x in the unit interval $[0, 1]$ such that

$$x\langle a, b\rangle + (1 - x)\langle a, c\rangle = 0 \quad . \qquad (iii)$$

Since G is convex, the quantity $xb + (1 - x)c$ belongs to G. Thus eq.(iii) shows that this element violates relation (i). Hence if (i) is true, (ii) must also hold. QED

This Lemma provides a simple test for whether or not a given n-tuple satisfies relation (10). Referring to eq.(9), once extremal elements $\widehat{f}$ and $\widehat{g}$ have been found which maximize $|\langle a, f\rangle|/|\langle a, g\rangle|$, then the n-tuple satisfies (10) if and only if the weighted sums $\langle a, \widehat{f}\rangle$ and $\langle a, \widehat{g}\rangle$ are of the same sign.

2.9.2 Exploiting Symmetry of the Response Set.

Relations (9) and (10) provide the basis for evaluating the expansion of the complete response set. Though the computational burden may seem to be prohibitive for a large number of detectors, it is completely computerizable and is considerably simplified by an important property of the minimizing n-tuple $a_1, \ldots, a_n$. Suppose two detectors, labelled j and k, have identical intrinsic characteristics and are located at symmetric positions relative to the sample. Then one can show [8] that $a_j = a_k$ in the minimizing n-tuple in eq.(9). In general, consider a configuration of n detectors in which there are m groups of symmetrically equivalent detectors. The coefficients a_i corresponding to each detector in a group of equivalent detectors may always be chosen to be equal. Furthermore, examination of eq.(9) shows that only the relative values of the coefficients are important. Hence the coefficients of one of the groups of symmetrically-equivalent detectors may automatically be equated to unity. It results that by exploiting the symmetry of the detector-sample configuration only $m - 1$ rather than n weights must be varied in the search for the minimizing n-tuple of coefficients. For example, if all the detectors are pair-wise symmetrically equivalent, then all the weights may be equated to unity, and no search on the set of n-tuples need be performed.

2.9.3 Summary of the Algorithm for Evaluating the Relative Mass Resolution

The algorithm for evaluating the relative mass resolution is summarized by relations (9) and (10). The algorithm requires a double iteration, with one iteration nested in the other. The "outer" iteration involves a search on the set of allowable n-tuples $a_1, \ldots, a_n$ until the minimum value of the ratio $\langle a, f \rangle / \langle a, g \rangle$ is found. The "inner" iteration requires a search on the point-source response set F for the minimum and maximum values of the weighted sum $\langle a, f \rangle$.

The following search procedure is recommended. An initial choice of the n-tuple is adopted; typically $a_1 = \cdots = a_n = 1$. Then elements u and v of F are sought which satisfy

$$\langle a, u \rangle = \underset{f \in F}{\text{minimum}} \langle a, f \rangle \qquad (2.9.11)$$

$$\langle a, v \rangle = \underset{f \in F}{\text{maximum}} \langle a, f \rangle \qquad (2.9.12)$$

This search is the "inner" iteration. Note that we have not included absolute value signs in eqs.(11) and (12). Having found u and v, the signs of the weighted sums are compared to verify that an allowed n-tuple has been adopted. The signs must agree.

Now the common coefficient of one of the groups of symmetrically-equivalent detectors is repeatedly altered while the other coefficients are held constant, and new values of u and v are obtained for each new n-tuple. This is continued until a local minimum is reached in the ratio

$$p(a_1, \ldots, a_n) = \left(\frac{\langle a, v \rangle}{\langle a, u \rangle} \right)^k \qquad (2.9.13)$$

where $k = 1$ if the weighted sums are positive; $k = -1$ if the weighted sums are negative [18]. Then the altered coefficient is held constant at its last value, and the next coefficient repeatedly

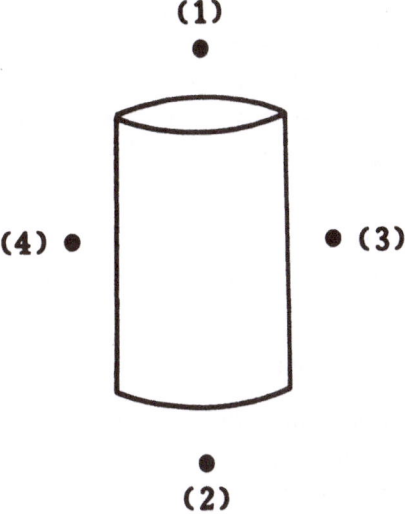

(1)

(4)   **(3)**

(2)

Figure 2.10.1 Deployment of four detectors around a cylindrical sample.

altered until a new minimum is again achieved, and so on for all the weights which must be varied. This successive variation of the components of the n-tuple constitutes the "outer" iteration. The whole procedure has to be repeated a number of times depending on the accuracy required.

2.10 EXAMPLE: Pu ASSAY WITH FOUR DETECTORS

In this section we shall illustrate the operation of the algorithm for evaluation of the relative mass resolution of a multi-detector system. We shall continue with the example initiated in section 2.6, and consider four identical detectors located around the cylindrical sample. All four detectors are 40 cm from the sample center. The response function for each detector is given by eq.(2.6.1). The arrangement of the four detectors is shown schematically in figure 2.10.1. Detectors numbered 1 and 2 are positioned on the rotational symmetry axis of the sample, one above and one below the sample. Detectors 3 and 4 are on the sample midplane separated by 180 degrees.

Detectors 1 and 2 have the same intrinsic characteristics and are symmetrically located with respect to the sample; the same is true for detectors 3 and 4. From the discussion of section 2.9.2 we conclude that the quartet of coefficients which minimize $p(a_1, \ldots, a_n)$ in eq.(2.9.13) will be of the form

$$a_1 = a_2 \quad ; \quad a_3 = a_4 \quad .$$

Since only the relative values of the coefficients are important, we may choose

$$a_3 = a_4 = 1 \quad .$$

Thus the "outer" iteration which searches for the quartet of coefficients which minimizes $p(a_1, \ldots, a_4)$ need search on a single dimension only: the value assigned to a_1 and a_2.

We begin by choosing $a_1 = a_2 = 1$. Now we enter the "inner" iteration and search for the minimum and maximum weighted sums, as indicated in eqs.(2.9.11) and (2.9.12). Having found the values of the extremal weighted sums, we advance to the next step of the "outer" iteration: alteration of the value of a_1 and a_2. Again maximum and minimum weighted sums are sought. In figure 2.10.2 is shown the value of the extremal sums, versus the value assigned to a_1 and a_2. The weighted sums are normalized to the value of the maximum sum at $a_1 = 1$. From the results of this figure we are able to calculate the ratio, p, of the maximum to minimum weighted sum, as in eq.(2.9.13). This is shown in figure 2.10.3. The minimum value of p is 2.88, occurring at $a_1 = a_2 = 0.094$. Thus the relative mass resolution of this four-detector arrangement is 2.88. That is, every spatial distribution in the sample of u grams of material is distinguishable from every spatial distribution of v grams of material if and only if

$$v > 2.88u$$

We note that this is appreciably better than the 1-detector and 2-detector configurations. With the detectors located on the sample midplane, 40 cm from the sample center, the 1- and 2-detector

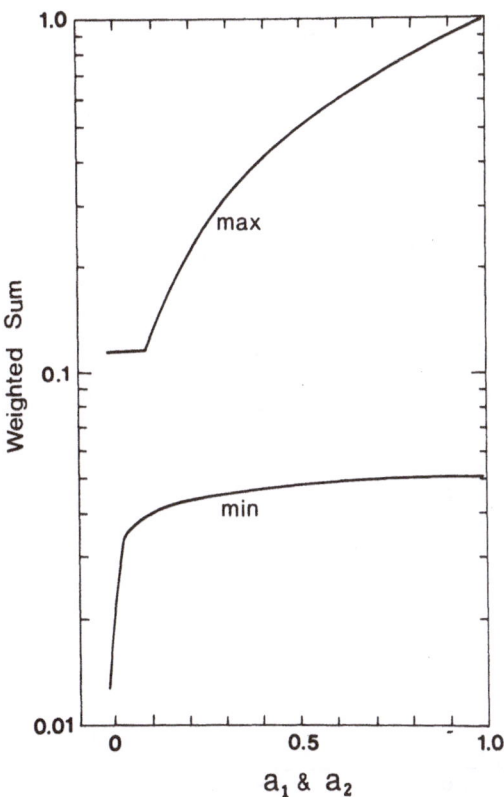

Figure 2.10.2 Maximum and minimum weighted sums versus the coefficients for detectors 1 and 2. The coefficients for detectors 3 and 4 equal unity.

configurations showed relative mass resolutions of 38.9 and 5.1, respectively.

Upon comparing figures 2.10.2 and 2.10.3 one notes that the minimum of $p(a_1, \ldots, a_4)$ occurs in an interval of a_1 in which both the minimum and maximum weighted sums show sharp change in slope. This change in slope is associated with a change in the point-source locations at which the extremal weighted sums occur. When a_1 and a_2 are much less than a_3 and a_4, the point-source location yielding the maximum weighted sum is on the sample midplane, at the sample surface just opposite one of the midplane

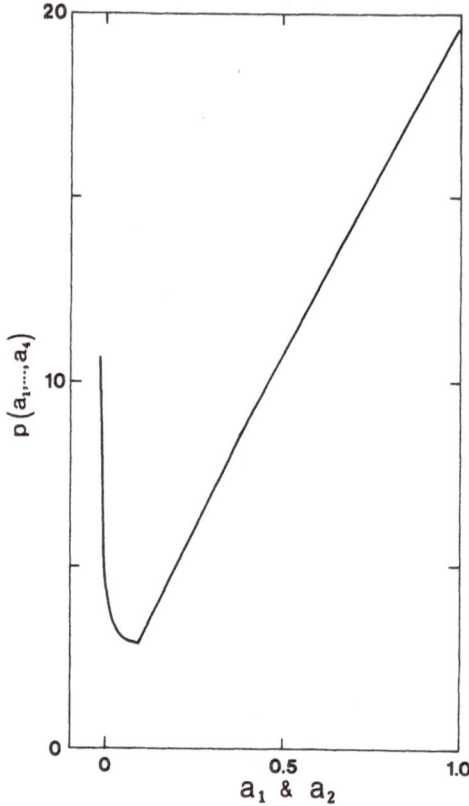

Figure 2.10.3   Ratio $p(a_1,...,a_4)$ of the maximum to the minimum weighted sums versus the coefficients for detectors 1 and 2. The co-efficients for detectors 3 and 4 equal unity.

detectors. When a_1 and a_2 are much greater than a_3 and a_4, the position yielding the maximum weighted sum is on the symmetry axis at the top or bottom of the sample. A very sharp transition between these two point-source locations occurs for $a_3 = a_4 = 1$ and $a_1 = a_2 = 0.095$.

The point-source location yielding the minimum value of the weighted sum also shows a transition, though less sharp than the transition of the position of the maximum sum. For a_1 and a_2 much greater than for a_3 and a_4, the position of the minimum is on the sample midplane at the sample surface. A transition occurs as a_1 and a_2 decrease, so that for $a_3 = a_4 = 1$ and $0.6 \leq a_1, a_2 \leq 1.0$, the position of the minimum sum is on the rotational symmetry axis at the very center of the sample. As a_1 and a_2 decrease further, the position of the minimum moves along the rotational symmetry axis toward the top, or bottom, of the sample.

It is sometimes possible to expedite the outer iteration — the search for the minimum value of $p(a_1, \ldots, a_n)$ — by recognizing that the minimum is likely to occur at or near a transition in the extremal point-source location. Also the inner iteration — the search for the extremal values of the weighted sum — may be quickened by noting that the extremal point-source locations usually remain constant over a considerable range of values of the coefficients. Care must be taken, however, because the transition of one or both of the extremal point-source positions may occur very abruptly.

2.11 THE INCLUSION OF STATISTICAL UNCERTAINTY

The concept of expansion which we have described in the previous sections expresses the relative mass resolution of the assay system in the absence of statistical uncertainty of measurement [19]. In order for this parameter to be a useful design tool, it is of prime importance to include consideration of the statistical uncertainty of the measurement. However, this must be achieved without unduly complicating the evaluation of the relative mass resolution.

2.11.1 Single-Detector Systems

To introduce the procedure for including statistical uncertainty in the evaluation of the relative mass resolution, we shall first consider a single-detector assay system. Let u and v be the minimum and maximum responses (count rates) obtainable from a unit source concentrated at a single point in the sample. The convexity theorem and the linearity assumptions assert that mtu and mtv

are the least and greatest number of counts obtainable from any spatial distribution of m grams of source material measured for a duration t. Let $\sigma(x)$ represent the statistical standard deviation of the counts x. For Poisson statistics [20] we have

$$\sigma(x) = \sqrt{x} \tag{2.11.1}$$

It is evident from our discussion of the concept of expansion that in order for any spatial distribution of $m+d$ grams to be distinguishable from any spatial distribution of m grams, we require that the smallest response obtainable from $m + d$ grams must exceed the greatest response obtainable from m grams. The response from $m+d$ grams may, in light of statistical uncertainty, be as small as $(m + d)tu - b\sigma\big((m + d)tu\big)$, where b is a positive quantity which determines the level of statistical confidence required. Likewise, the response from m grams may be as great as $mtv + b\sigma(mtv)$. Thus any spatial distribution of $m + d$ grams will be distinguishable to the desired degree of statistical confidence from any spatial distribution of m grams if and only if

$$(m + d)tu - b\sqrt{(m + d)tu} > mtv + b\sqrt{mtv} \tag{2.11.2}$$

The smallest ratio $(m + d)/m$ which satisfies this relation when expressed as an equality is the relative mass resolution with statistical uncertainty:

$$\frac{m + d}{m} = \left(\sqrt{z} + \frac{b}{\sqrt{mtu}}\right)^2 \tag{2.11.3}$$

where $z = v/u$ is the relative mass resolution in the absence of statistical uncertainty. The expression on the righthand side of eq.(3) is the relative mass resolution with statistical uncertainty, henceforth denoted Z.

Let us consider the one-detector example discussed in section 2.6, with the detector on the midplane of the sample. We concluded that, in the absence of statistical uncertainty, the relative mass resolution improves (decreases) as the detector is removed from the sample (see figure 2.6.3), and approaches an asymptotic limit at infinite separation (eq.(2.6.5)). That is, z in eq.(3) decreases with increasing detector-sample separation. However we should expect the contribution of the statistical uncertainty to increase as the detector is removed from the sample due to the decreasing count rate. This is demonstrated by the second term on the righthand side of eq.(3). For close placement of the detector the statistical uncertainty may indeed be negligible, and the overall mass resolution is due entirely to spatial uncertainty as embodied in z. However, as the separation increases, the contribution of the statistical uncertainty to the relative mass resolution increases without bound. From these considerations we should expect the mass resolution with statistical uncertainty to be a convex function, showing a minimum (optimum) value for some intermediate sample-detector separation. This is shown schematically in figure 2.11.1. We will observe this phenomenon in all our subsequent examples. Indeed much of the task of the assay-system designer is to find the best compromise between the restraints imposed by the spatial uncertainty on the one hand, and those imposed by the statistical uncertainty on the other.

2.11.2 Multi-Detector Systems

To generalize these results to the case of n detectors we proceed as follows. Let u_i and v_i, $i = 1, \ldots, n$, be the responses of the n detectors which, respectively, minimize and maximize the weighted sums in eqs.(2.9.11) and (2.9.12). Also let $h_1, \ldots, h_n$ be the n-tuple of coefficients which minimizes the ratio $|\langle a, v \rangle / \langle a, u \rangle|$ in eq.(2.9.8). Thus the relative mass resolution (without inclusion of statistical uncertainty) is

$$z = \frac{\langle h, v \rangle}{\langle h, u \rangle} \qquad (2.11.4)$$

(We are assuming for convenience that the weighted sums are positive). In other words, quantities m and $m + d$ are distinguishable

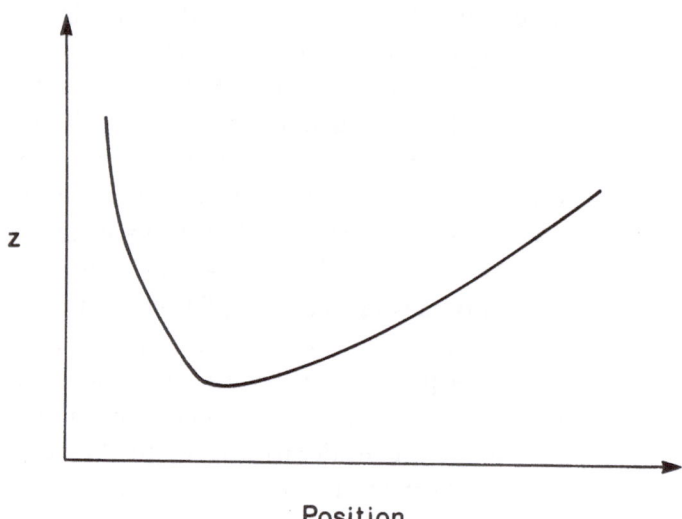

Position

Figure 2.11.1 Schematic representation of the relative mass resolution
with statistical uncertainty versus the sample-detector separation.

by a measurement of arbitrary duration t if and only if

$$\langle h, (m+d)tu \rangle > \langle h, mtv \rangle \qquad (2.11.5)$$

To include statistical uncertainty in relation (5), let $\sigma(x)$ be the
statistical standard deviation of an arbitrary quantity x. Thus m
grams are distinguishable from $m+d$ grams, to a specified level of
statistical confidence, if the smallest likely value of the left-hand
side of ineq.(5) exceeds the largest likely value of the right-hand
side. That is if:

$$\langle h, (m+d)tu \rangle - b\sigma\big(\langle h, (m+d)tu \rangle\big)$$
$$> \langle h, mtv \rangle + b\sigma\big(\langle h, mtv \rangle\big) \qquad (2.11.6)$$

where b is a positive number indicating the level of statistical confidence required. The statistical standard deviation of a weighted
sum of measurements is

$$\sigma\big(\langle h\,, mtf\rangle\big) = \left(\sum_{i,j} h_i h_j \mathrm{cov}(mtf_i, mtf_j)\right)^{1/2} \qquad (2.11.7)$$

where $\mathrm{cov}(x,y)$ is the covariance of x and y.

If the measurements are all independent Poisson random variables, then eq.(7) becomes

$$\sigma\big(\langle h\,, mtf\rangle\big) = \left(\sum_{i} h_i^2\, mtf_i\right)^{1/2} \qquad (2.11.8)$$

Also, ineq.(6) becomes an implicit relation in $(m+d)/d$, whose solution is the relative mass resolution with spatial and statistical uncertainty:

$$Z = \frac{m+d}{m} \qquad (2.11.9)$$

$$= \left((z + V^2 + T)^{1/2} + V\right)^2 \qquad (2.11.10)$$

where z is given by eq.(4) and

$$T = \frac{b\sigma\,(\langle h\,, v\rangle)}{\langle h\,, u\rangle\sqrt{mt}} \qquad (2.11.11)$$

$$V = \frac{b\sigma\,(\langle h\,, u\rangle)}{2\langle h\,, u\rangle\sqrt{mt}} \qquad (2.11.12)$$

where $\sigma(x)$ is the statistical standard deviation for independent Poisson measurements, given by eq.(8).

2.12 EXAMPLE: RADIOACTIVE WASTE ASSAY

In this section we shall illustrate some aspects of the design of an instrumental system for the assay of plutonium in a polyethylene matrix. The properties of the cylindrical sample are those given in the example in section 2.6.

The design decisions which one confronts are:

1. What type of radiation should be employed? One can passively measure the 384 keV gamma complex and/or other gamma radiations [21]. Alternatively, passive neutron techniques [22] or neutron activation techniques [23] could be used. Closely related to this is the question of what type of detector to use.

2. How many detectors should be used, and how should they be deployed around the sample?

3. Should the detectors and/or the sample be stationary or moving?

4. What should be the duration of the measurement?

5. In some situations one is free to choose the size and shape of the sample [24].

In our example we shall examine passive gamma assay with sodium iodide detectors. We will concentrate on questions 2 and 3.

The response function of the detectors to be used is

$$f(d, D) = \frac{gspa}{4\pi D^2} e^{-\mu d} \qquad (2.12.1)$$

where g is the fractional yield of the relevant radiation, s is the specific activity of the source material (counts/g sec), p is the intrinsic peak efficiency of the detector and a is the effective detector area (cm^2). D is the total distance from the point-source to the detector, d is the distance through the sample along the same line, and μ is the linear absorption coefficient of the sample matrix. For the particular source material and NaI crystal we are using the constants in eq.(1) are

$$\frac{gspa}{4\pi} = 1.367 \times 10^3 \qquad \text{counts/g sec}$$

The value of μ is 0.10 cm^{-1}.

We shall consider a measurement duration of 10 seconds, and a total source mass of 1 gram of plutonium. The statistical uncertainty is evaluated at two standard deviations. That is, $b = 2$ in eq.(2.11.6).

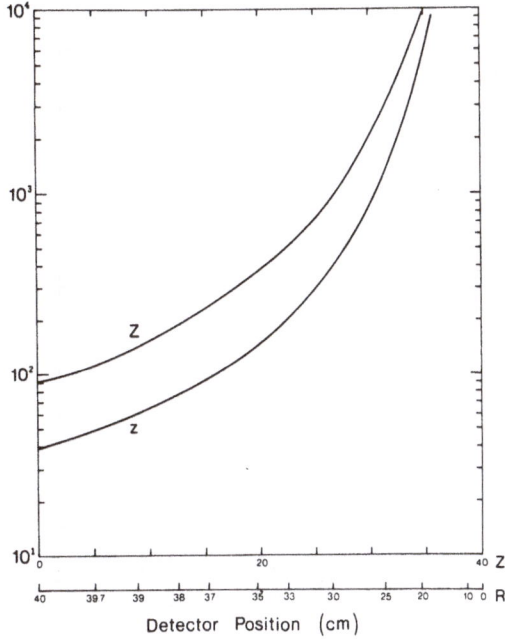

Figure 2.12.1 Relative mass resolution with statistical uncertainty (Z) and without statistical uncertainty (z) versus the position of a single detector on a quarter circular arc of 40 cm radius. Stationary sample and detector.

We shall first consider stationary designs. To begin to get a feeling for the system we are designing, let us consider a range of one-detector alternatives. We start by calculating the relative mass resolution for a single detector at various positions along a quarter-circular arc of radius 40 cm around the center of the sample, as in figure 2.6.1. In figure 2.12.1 we show the relative mass resolution with statistical uncertainty (Z) and without statistical uncertainty (z). The best resolution, which is indeed quite poor, is obtained with the detector on the midplane, as we observed in section 2.6. The relative mass resolution with statistical uncertainty is 89.1, and without statistical uncertainty is 38.9

In figure 2.12.2 we show the relative mass resolution of a single midplane detector, as a function of the sample-detector separa-

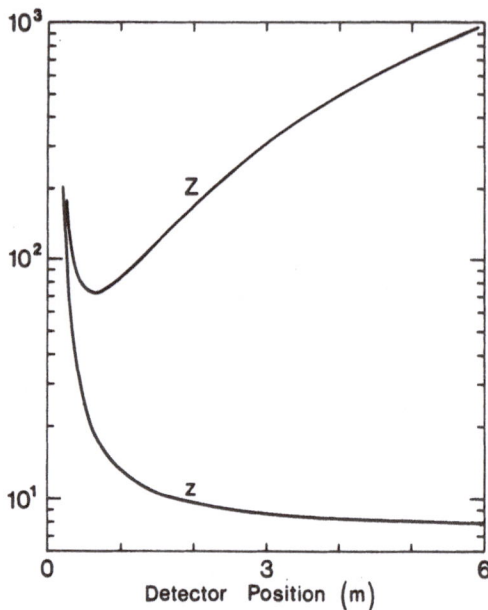

Figure 2.12.2   Relative mass resolution with statistical uncertainty (Z) and without statistical uncertainty (z) versus the distance from the sample-center of a single midplane detector. Stationary sample and detector.

tion. The relative mass resolution without statistical uncertainty, z, asymptotically approaches a limit as the detector is removed, as we have seen in previous examples. The resolution with statistical uncertainty, Z, shows a minimum of 70.3 at 60 cm. At closer separations Z approaches z, since the dominant contribution to the uncertainty is spatial: the random spatial distribution of the source material. At greater separations Z increases without bound since the detector response decreases, causing increasing relative statistical uncertainty. The optimum sample-detector separation occurring at 60 cm is a result of the compromise between these conflicting tendencies.

We are now ready to explore the characteristics of stationary multi-detector design options. Let us first consider symmetrical arrangements of detectors on the sample midplane. For n detectors, neighboring detectors are separated by $360/n$ degrees. Calculation of the mass resolution is particularly simple in this case since every pair of detectors are symmetrically equivalent with respect to the sample. From our discussion in section 2.9.2 we are able to conclude that the n-tuple of coefficients which minimize $p(a_1, \ldots, a_n)$ in eq.(2.9.13) are all equal to unity. In Table 2.12.1 we present the relative mass resolution for symmetric midplane detectors. We see that two detectors provide much better resolution than one, and that four detectors are somewhat better than two. For more than four detectors the utility of the marginal detector is small.

Table 2.12.1

Relative Mass Resolution
for Equally Spaced Midplane Detectors
100 cm from the Sample Center

n	1	2	4	8	16
z	12.8	2.52	2.30	2.19	2.17
Z	82.3	13.8	9.13	6.47	5.24

It is now appropriate to explore various other configurations of four detectors around the sample. Figure 2.12.3 shows the relative mass resolution with and without statistical uncertainty for four equally spaced midplane detectors, as a function of the detector distance from the sample center. As is to be expected, z decreases to an asymptot, while Z shows a minimum value of 7.18 at 60 cm. Figure 2.12.4 shows the same calculation for a different detector deployment: two detectors on the midplane separated by 180 degrees, and two detectors on the rotational symmetry axis one above and one below the sample. The minimum value of the relative mass resolution is 6.37, and occurs at 40 cm.

From this analysis we may conclude that the most promising

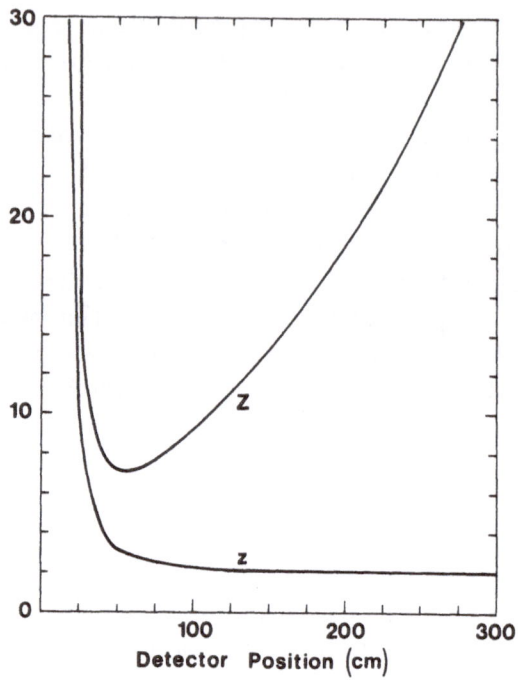

Figure 2.12.3 Relative mass resolution with statistical uncertainty (Z) and without statistical uncertainty (z) versus distance from the sample-center for four symmetric midplane detectors. Stationary sample and detectors.

stationary configuration of NaI detectors calls for two axial detectors and two midplane detectors, with all detectors 40 cm from the sample center. In a more extensive design analysis we may wish to examine placing the axial and midplane detectors at different distances. It may also be feasible to consider greater measurement duration. This will influence the statistical uncertainty and thus change the optimum distances.

We shall now address question 3, and explore the advantage of rotating the sample around its rotational symmetry axis. The sample will perform one complete revolution during the 10-second measurement [25].

Let us first consider a single midplane detector. In figure 2.12.5

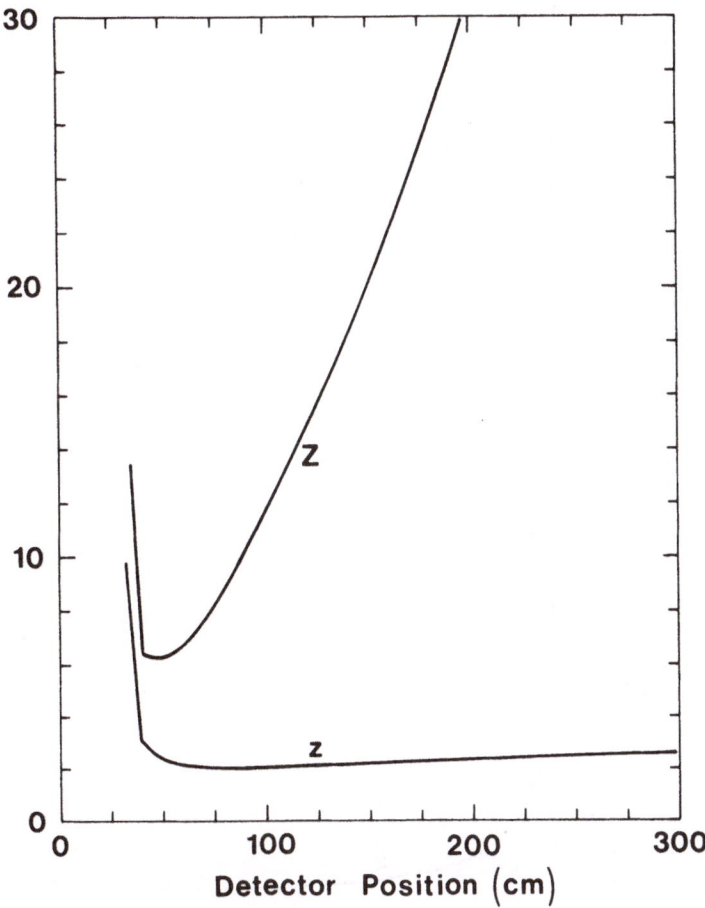

Figure 2.12.4 Relative mass resolution with statistical uncertainty (Z) and without statistical uncertainty (z) versus distance from the sample-center, for two midplane detectors and two axial detectors. Stationary sample and detectors.

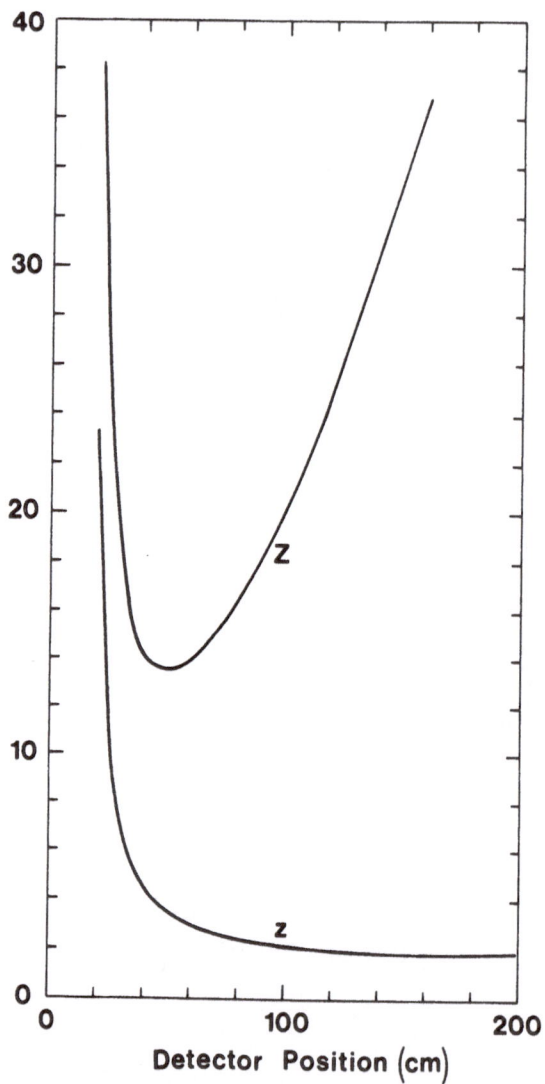

Figure 2.12.5 Relative mass resolution with statistical uncertainty (Z) and without statistical uncertainty (z) versus the distance from the sample-center of a single midplane detector. Rotating sample.

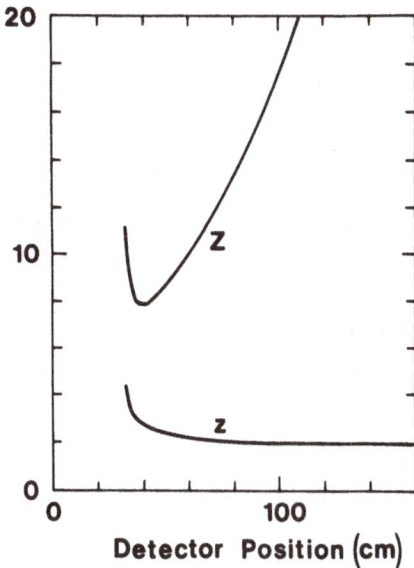

Figure 2.12.6 Relative mass resolution with statistical uncertainty (Z) and without statistical uncertainty (z) versus the distance from the sample-center of one midplane detector and two axial detectors. Rotating sample.

we plot the relative mass resolution with and without statistical uncertainty versus the radial position of the detector. As usual, z asymptotically approaches a limiting value, while Z shows a minimum of 13.5 at 50 cm. Rotation of the sample provides striking improvement, as we see by comparing these results with figure 2.12.2.

In figure 2.12.6 we examine the effect of employing two axial detectors in addition to the radial detector. In this case the minimum value of the relative mass resolution with statistical uncertainty is 7.77 and occurs with all three detectors positioned 40 cm from the sample center.

Table 2.12.2

Summary of Design Analysis

Sample	Detector Locations	Z†	z‡
stationary	one midplane	70.40	20.40
rotating	one midplane	13.50	3.48
stationary	two midplane and two axial	7.18	2.91
rotating	one midplane and two axial	7.77	2.63

† Optimum value.

‡ Value taken at optimal sample-detector separation.

In Table 2.12.2 we summarize the results of our design analysis. From the first two lines we see that rotating the sample greatly improves the resolution capability of a single midplane detector. From the third and fourth lines we note that the addition of two axial detectors to the rotating-sample design results in further improvement, and that the three-detector rotating design is nearly as good as the stationary four-detector case. This analysis indicates that considerable advantage is to be gained by rotating the sample. However, our final judgement of the relative merits of these alternative designs must be deferred, as we shall explain in the next section.

2.13 SUMMARY

2.13.1 The Deterministic Measure of Performance

The major effort of this Chapter has been the development of an efficient computerizable algorithm for evaluating the deterministic measure of performance: the relative mass resolution without statistical uncertainty, z. This measure is defined so that, for quantities u and v of source material, any sample containing v grams is distinguishable from any sample containing u grams if and only if

$$v > zu$$

In other words, u and v grams are always distinguishable if and only if every spatial distribution of u grams within the sample yields a set of measurements which is distinct from the measurements resulting from every spatial distribution of v grams. This criterion is deterministic in the sense that no attempt is made to assess the probability of the different spatial distributions of the source material within the sample. Suppose v is sufficiently greater than u so that only very unusual and unlikely spatial distributions of v grams yield measurements obtainable also from u grams. In such a case we might in practice wish to consider the quantities u and v as distinguishable. However, according to the deterministic criterion, u and v are not distinguishable.

Even if z is appreciably greater than unity, it might in fact be that the practical resolution is much better, and that only very unlikely spatial distributions are "preventing" z from assuming a lower value. In the example summarized in Tables 2.12.1 and 2.12.2 we see that z seems invariably to take a value greater than 2. In most situations a relative mass resolution of 2 is intolerably large: 100% uncertainty. We may suspect (or hope) that in fact only very improbable spatial distributions of u grams yield measurements which could also be obtained from $2u$ grams. In other words, z may give a pessimistic evaluation of the resolution capability. Referring to the two rotating-sample designs summarized in Table 2.12.2, we see that the addition of two axial detectors has almost no effect on z. From the point of view of the deterministic criterion z these two designs are of equal quality. However we might expect that, for the three-detector design, the probability of encountering an ambiguous set of measurements is appreciably lower than for the one-detector design.

In Chapter 4 we shall begin our study of the probabilistic anal-

ysis of assay systems, by discussing the probabilistic interpretation
of measurement. This will lay the ground work for the probabilistic design analysis presented in Chapter 5.

2.13.2 Statistical Uncertainty

The second important concept introduced in this Chapter is
the extension of the deterministic measure of performance z to
include the statistical uncertainty of measurement. The statistical
uncertainty arises from randomness of the pulse-counting process,
and is important for measurements of short duration or for large
separation between sample and detector. We observed that the
statistical uncertainty can be mitigated by placing the detectors
close to the sample. On the other hand, the spatial uncertainty —
arising from random spatial distribution of the source material in
the sample — is reduced by removing the detectors from the sample. The design process is a search for the optimum compromise
between these conflicting tendencies.

NOTES

[1] The relative mass resolution is by no means the only concept
which may be used to define a measure of performance. In much
design work of nuclear assay systems the degree of "flatness" of
the response function has been used as a measure of performance.
An additional useful and widely employed design concept is the
resolving power. See:
1. A. Notea and Y. Segal, A General Approach to the Design of
Radiation Gauges, *Nucl. Tech.*, 14: 73–80 (1974).
2. Y. Segal, A. Notea and E. Segal, A Systematic Evaluation of
Nondestructive Testing Methods, in *Research Techniques in Nondestructive Testing*, Vol. III, R. S. Sharpe, ed., Academic Press,
1977.
3. N. Shenhav, Y. Segal and A. Notea, Application of Neutron
Count Moment Analysis Method to Passive Assay, *Nucl. Sci.
Eng.*, 80: 61–73 (1982).

[2] The concept of the response function does not require that the detector be localized at a single point. It is meaningful for well-type counters which entirely surround the sample and for calorimetric measurements. For an interesting exploitation of the sensitivity of a NaI well-type counter to the spatial distribution within the well of a gamma source see
1. F. Cejnar, L. Wilhelmova and I. Kovar, Determination of Radionuclide Two-phase Distribution by Well-Type NaI(Tl) Detectors, *Int. J. Appl. Rad. Isotopes*, 32: 1–4 (1981).
For a discussion of calorimetric measurements see
2. J. D. Nutter, F. A. O'Hara and W. W. Rodenburg, The Use of Calorimetry in Nuclear Materials Management, *Inst. Nucl. Mat. Mgt.*, 11-th annual conf., May 1970, Gatlinburg, Tenn., pp. 101–121.

[3] See G. Knoll, *Radiation Detection and Measurement*, John Wiley, 1979, for a discussion of mean free paths.

[4] The probability of shadowing can readily be evaluated analytically in certain circumstances. For a discussion of free paths in a Poisson ensemble of disks or spheres see:
W. Feller, *An Introduction to Probability Theory and its Applications*, Vol. II, John Wiley, 1971.

[5] Many examples may be found in
T. Gozani, *Active Nondestructive Assay of Nuclear Materials*, U.S. Natl. Tech. Inf. Serv., NUREG/CR-0602, 1981.

[6] Let M be a subset of a metric space R. A point x in R is called a *limit point* of M if every neighborhood of x contains an infinite number of elements of M. A set is *closed* if it contains all its limit points. Thus for example an interval and its endpoints constitute a closed set on the real line.

[7] A set is convex if the straight line joining any two points of the set is entirely in the set. Algebraically, this may be expressed by requiring that, for any x in the closed interval $[0, 1]$ and any elements g and h of the set, the quantity $xg + (1 - x)h$ is also an

element. For further discussion of basic properties of convex sets
see

1. A. N. Kolmogorov and S. V. Fomin, *Introductory Real Analysis*,
Dover, 1970.
2. P. Kelly and M. L. Weiss, *Geometry and Convexity*, Wiley
Interscience, 1979.
3. S. R. Lay, *Convex Sets and Their Applications*, John Wiley,
1982.

[8] Y. Ben-Haim and N. Shenhav, The Measurement of Spatially
Random Phenomena, *S.I.A.M. J. Appl. Math.*, 44: 1150–63
(1984).

[9] For simple sample shapes it is rarely necessary to evaluate the
response functions for a large number of point-source positions.
Usually the boundary of the response set can be constructed by
evaluating f_1 and f_2 along a simple path within the sample.

[10] We will show in section 2.9.2 that when the detectors are in
symmetric positions relative to the sample, the ray of maximum
expansion always passes through the symmetry axis of C.

[11] It is interesting to note that in this two-detector example the
expansion as a function sample-detector distance is not a monotonic function. Though it is bounded, it passes through a minimum which is somewhat less than its asymptotic value. For a
discussion of this see
Y. Ben-Haim and N. Shenhav, The Design of Nondestructive Assay Systems, *Int. J. Appl. Rad. Isotopes*, 34: 1291–1299 (1983).

[12] N. E. Holden and M. S. Zucker, Plutonium Nuclide Nuclear
Data Standards of Safeguards Interest, *6-th Ann. Symp. Safeguards and Nuclear Mat. Mgt.*, ESARDA, Venice, May 1984.

[13] 1. T. Gozani and D. G. Costello, Isotopic Source Assay
System for Nuclear Materials, *Trans. Am. Nuc. Soc.*, 13:746–7
(1970).
2. T. Gozani and D. G. Costello, On the Reduction of Gamma
Self-Shielding Using the Fission Multiplicity Detector, *Trans. Am.*

Nuc. Soc., 14:809–10 (1971).
3. M. S. Zucker and E. Karpf, Absolute Calibration of Spontaneous Fission Neutron Sources, 6-th Ann. Symp. Safeguards and Nuclear Mat. Mgt., ESARDA, Venice, May 1984.

[14] See ref. [5] and
T. Gozani, Fission Multiplicity Detectors, Trans. Am. Nuc. Soc., 24:127–8 (1976).

[15] I. S. Panasyuk, On Measurements of Disintegration Rates of Radioactive Sources by Coincidence Methods, Nucl. Inst. Meth., 31:314–6 (1964).

[16] T. Gozani, Reduction of Geometry and Matrix Effects by Fast Coincidence Non-Destructive Assay, Proc. Inst. Nucl. Mat. Mgt., 12: 195–206 (1983).

[17] The material in this section is discussed from various points of view in ref. [8] and in
N. Shenhav and Y. Ben-Haim, A General Method for the Design of Nondestructive Assay Systems, Nucl. Sci. Eng., 88: 173–83 (1984).

[18] If the weighted sums are positive, then the greatest sum is also the greatest sum in absolute value. In this case eq.(2.9.9) could be written without the absolute-value signs, and in eq.(2.9.13) k is set equal to unity. If the weighted sums are negative, then sum $\langle a, v \rangle$ will be of least absolute value and the sum $\langle a, u \rangle$ will be of greatest absolute value. That is (for negative weighted sums)

$$\langle a, u \rangle = \operatorname*{minimum}_{f \in F} \langle a, f \rangle = \operatorname*{maximum}_{f \in F} |\langle a, f \rangle|$$

$$\langle a, v \rangle = \operatorname*{maximum}_{f \in F} \langle a, f \rangle = \operatorname*{minimum}_{f \in F} |\langle a, f \rangle|$$

The exponent k in eq.(2.9.13) is set to -1 to perform the required inversion. In eq.(2.9.9) we avoided this complication by the use of the absolute value; this was allowed by specifying that the n-tuple satisfies relation (2.9.10). From the operational point of view we prefer the form of eq.(2.9.13) since we do not know whether or not the n-tuple satisfies (2.9.10) until we compare the signs of the maximum and minimum weighted sums. It is therefore important to stress that the weighted sums are extremal in actual, not absolute, value.

[19] We are referring to the statistical uncertainty of the individual measurement, arising from randomness of the pulse-counting process. This is to be distinguished from the propagation of error when a series of measurements are combined in some way. For a discussion of the propagation of statistical uncertainty in nuclear material management see:
1. L. D. Y. Ong *et al*, A Look at SNM Measurement Errors Based on Safeguards Inventory Verification Duplicate Samples, *Inst. Nucl. Mat. Mgt.* 11-th annual conf., May 1970, Gatlinburg, Tenn., pp. 177–89.
2. J. Lieblein, Generalized Propagation of Error Using a New Approach, *ibid*, pp. 190–211.
3. T. E. Sampson and R. Gunnink, Propagation of Errors in the Measurement of Plutonium Isotope Composition by Gamma-Ray Spectroscopy, *J. Inst. Nuclear Mat. Mgt.*, 12:39–48 (1983).

[20] For thorough discussion of the fundamentals of the Poisson process see
1. W. F. Feller, *An Introduction to Probability Theory and its Applications*, Vol. I, John Wiley, 1968.
2. A. Papoulis, *Probability, Random Variables and Stochastic Processes*, McGraw-Hill, 1965.
For a discussion of how the Poisson process arises in electronic systems see
3. S. O. Rice, Mathematical Analysis of Random Noise, in *Noise and Stochastic Processes*,ed. by N. Wax, Dover Pub., 1954.
For a discussion of how Poisson processes arise in neutronic systems see
4. M. M. R. Williams, *Random Processes in Nuclear Reactors*, Pergamon Press, 1974.

[21] 1. J. W. Kormuth, J. F. Gettings, D. B. James, Nondestructive Assay of Plutonium Fuel Plates, *Inst. Nucl. Mat. Mgt.* 11-th annual conf., May 1970, Gatlinburg, Tenn., pp. 266–79.

2. T. W. Packer, R. Howsley and E. W. Lees, Measurement of the Enrichment of Uranium Deposits in Centrifuge Pipework, *6-th Ann. Symp. Safeguards and Nuclear Mat. Mgt.*, ESARDA, Venice, May 1984.

3. T. Dragnev, Spectrometric Gamma Absorption Measurements, *ibid.*

4. N. Hayano, *et al*, Evaluation of Leached Hull Monitoring System as the Tokai Reprocessing Plant, *ibid.*

[22] For a good discussion of several passive neutron techniques see

1. J. L. Parker *et al*, Passive Assay — Innovations and Applications, *Inst. Nucl. Mat. Mgt.*, 12-th ann. conf., June 1971, Palm Beach, Fla, pp. 514–47.

2. M. Despres *et al*, Analysis Code for Plutonium Isotopic-Composition Measurements, *Int. J. Appl. Rad. Isotopes.*, 34: 525–31 (1983).

3. L. Bondar *et al*, Determination of Plutonium in Mixed Oxide Fabrication Plant Wastes by Passive Neutron Assay, *6-th Ann. Symp. Safeguards and Nuclear Mat. Mgt.*, ESARDA, Venice, May 1984.

For an example of a passive neutron technique which discusses the effect of spatial distribution of the assayed material in the sample, see

4. A. Notea, Y. Segal, A. Bar-Ilan, A. Knoll and N. Shenhav, Interpretational Models of Passive Gamma and Neutron Assay Systems, in *Monitoring of Pu-Contaminated Waste*, EUR 6629 EN Sept. 1979, pp 363–81.

5. A. Knoll, A. Notea, Y. Segal, Probabilistic Interpretation of Nuclear Waste by Passive Gamma Technique, *Nucl. Tech.*, 56: 351–60 (1982).

6. M. S. Zucker and E. V. Weinstock, Passive Measurements of Waste in 55 Gallon Drums and Conventional Scrap in 2 Liter Bot-

tles, *Inst. Nucl. Mat. Mgt.*, 11-th annual conf., May 1970, Gatlinburg, Tenn., pp. 281–315.

[23] 1. R. A. Harlan *et al*, Neutron Interrogation of Low-Specific-Activity Wastes with a Reactor, *Inst. Nucl. Mat. Mgt.*, 12-th annual conf., June 1971, Palm Beach, Fla., pp. 373–98.
2. M. M. Thorpe, *et al*, Active Assay of Fissionable Materials, *Inst. Nucl. Mat. Mgt.*, 12-th annual conf., June 1971, Palm Beach, Fla., pp. 631–47.
3. R. A. Harlan et al, Operational Assay for Fissile Material in Crated Nuclear Energy Wastes, *6-th Ann. Symp. Safeguards and Nuclear Mat. Mgt.*, ESARDA, Venice, May 1984.
4. P. Dell'Oro *et al*, Field Experience on Non-Destructive Assay of Low-Enrichment Uranium by Means of Photoneutron Interrogation, *ibid*.

[24] For an example see the reference in note [17].

[25] For an interesting design which exploits rotation of the sample and collimation of the detector see
H. O. Menlove *et al*, Neutron Interrogation Techniques for Fissionable Material Assay, *Inst. Nucl. Mat. Mgt. 11-th Annual Conf.*, May 1970, Gatlinburg, Tenn., pp.316–47.

CHAPTER 3

DETERMINISTIC DESIGN II:
GENERAL FORMULATION

3.1 MOTIVATION

The previous Chapter was devoted to developing the conceptual foundations of the deterministic design-analysis. The concept of relative mass resolution was introduced as a deterministic measure of performance. The convexity theorem established a simple analytic relation between the point-source response set and the complete response set. This Theorem leads to the conclusion that the relative mass resolution is precisely equal to the expansion of the complete response set. Furthermore, an efficient computerizable min-max algorithm was established which enables evaluation of the expansion of the complete response set, while requiring explicit knowledge only of the point-source response set. Finally, the concept of relative mass resolution was extended to include the statistical uncertainty of the measurement.

The present Chapter is directed toward generalizing the concepts and techniques introduced previously. To do so it will be necessary to establish a clear mathematical foundation, rather than to rely on the intuitive approach used up to now. In return we shall attain a vast extension of the realm of assay problems which can be studied.

The assay applications which have occupied us up to now have involved the measurement of the total quantity of a material randomly distributed in the spatial domain of the sample. By considering the spatial distribution of material as an unnormalized probability density, such a measurement may be viewed as determination of the zeroth moment of the spatial distribution. Important assay applications arise in which the aim is to measure the first moment, or average position, of the analyte. For example, a standard technique for measurement of the flow rate of a fluid involves injection (or activation) of an isotope at one point in a flow channel, and measurement of the time at which the isotope passes a different point downstream [1 − 3]. Due to dispersive effects in the flowing fluid, the isotope passes the detection point as a spatial distribution whose shape may be poorly known. Thus the problem of determining when the isotope passes the detection point requires measuring where the material is as a function of time. This assay is in fact the assay of the first moment of the time-varying spatial distribution of the isotope. In other applications one wishes to measure the spatial dispersion of the analyte, as represented by the second or perhaps higher moments of the spatial distribution. In this Chapter we shall extend our previous study, and develop a measure-of-performance for the assay of an arbitrary moment of the analyte spatial distribution.

In our previous study we assumed that a given quantity of analyte can assume any conceivable spatial distribution. For a wide range of applications this is in fact a reasonable hypothesis. We continue with this assumption in section 3.2, and extend the concept of relative resolution to the assay of any moment. Having established this foundation we continue in section 3.3 by imposing constraints on the set of allowed spatial distributions of the analyte. This opens up a wide range of applications, several of which we examine in sections 3.4 to 3.9. In section 3.10 we introduce a further generalization, and allow the random variation of parameters other than the spatial distribution of the analyte. The most important example of a variable auxiliary parameter is random variation of the non-analyte matrix material. This is studied in sections 3.11 and 3.12. Finally in section 3.13 we consider the assay of arbitrary moments of constrained time varying distributions. The flow-rate measurement example is discussed in section 3.14.

3.2 UNCONSTRAINED SPATIAL DISTRIBUTIONS

In this section we shall establish a mathematical basis for the results which were intuitively derived in Chapter 2. Our notation remains as before. Let x be a vector variable representing position within the sample, and let X be the set of all vector-values which x may obtain. A specific assay system comprises n detectors. The j-th detector is characterized by a response function, $f_j(x)$, which represents the time-averaged response of the detector to a unit mass of assayed material concentrated at point x in the container. The response functions are non-negative real functions for x in X. It is convenient to let $f(x)$ represent the vector response function defined as

$$f(x) = \big(f_1(x), \ldots, f_n(x)\big) \qquad .$$

Let $r(x)$ be a non-negative real scalar function defined on X representing the density of analyte material at point x. We shall assume, as in Chapter 2, that the assay system displays linearity in mass. That is, the measured response to the spatial distribution $r(x)$ is

$$c(r) = \int r(x) f(x) \, dx \qquad (3.2.1)$$

Here, as throughout the Chapter, the range of integration is the set X unless otherwise specified.

We now wish to define the ordinary moments of the spatial distribution r in the domain X of the container. If x is a three-dimensional variable, we shall be interested in moments of the form

$$m(ijk) = \int x_1^i \, x_2^j \, x_3^k \, r(x) \, dx \qquad (3.2.2)$$

A more compact and general formalism is useful, so let h be a vector of non-negative integers where the dimensions of h and of x are equal. The i-th element of h is the exponent of x_i in eq.(2), and the product in the integrand of eq.(2) is represented succinctly as

$$x_1^i \, x_2^j \, x_3^k = x[h] \tag{3.2.3}$$

Thus eq.(2) becomes

$$m(h) = \int x[h] r(x) \, dx \tag{3.2.4}$$

and $m(h)$ is referred to as the h-moment of r. Usually we are interested in low-order moments, so the sum of the elements of h is generally a small integer.

Eq.(1) shows that the measurement vector c may be viewed as a measurement of the zeroth moment of the distribution r, as distorted by the point-source response function f. However, if X does not contain the origin, we may write c as

$$c(r) = \int x[h] r(x) \frac{f(x)}{x[h]} \, dx \tag{3.2.5}$$

which shows c as a distorted measurement of the h-moment of $r(x)$.

Let $R(h, u)$ represent the set of all functions integrable on X whose h-moment equals u. The set $R(h, u)$ is convex. To see this, let r and s be any elements in $R(h, u)$. Then for any $0 < y < 1$, $yr + (1 - y)s$ is integrable and its h-moment equals u. Thus $yr + (1 - y)s$ belongs to $R(h, u)$. We now define an important property.

DEFINITION 1 Let $\{A(u), u > 0\}$ be a class of sets defined for a positive real parameter u. Let $uA(1)$ be the set obtained from $A(1)$ by multiplying each element of $A(1)$ by u. This class of sets has the *property of proportionality* if

$$A(u) = uA(1)$$

It is evident from the definition of the h-moment in eq.(4) that the class of sets $\{R(h, u), \ u > 0\}$ has the property of proportionality.

The set of all possible vector values which the measurement vector c may obtain in response to spatial distributions whose h-moment equals u, is given by

$$C(h, u) = \left\{ c : c = \int r(x) f(x)\, dx \quad \text{for all} \quad r \in R(h, u) \right\}$$
(3.2.6)

This set is called a *complete response set* for the h-moment. It is readily shown that $C(h, u)$ is a convex set since $R(h, u)$ is convex. Furthermore, the class of sets $\{C(h, u),\ u > 0\}$ has the property of proportionality.

We noted in connection with eq.(5) that the vector c may be viewed as a measurement of the h-moment of r, distorted by $f(x)/x[h]$. This leads us to a generalization of the point-source response set F introduced in Chapter 2. The point-source response set for the h-moment, $F(h)$, is the set of all vector values which $f(x)/x[h]$ may obtain. That is

$$F(h) = \left\{ g : g = \frac{f(x)}{x[h]} \quad \text{for all} \quad x \in X \right\}$$
(3.2.7)

According to this definition, the point source response set F defined in Chapter 2 is $F(0)$. The importance of the set $F(h)$ is revealed by the following generalization of the convexity theorem.

THEOREM 1 If X does not contain the coordinate origin, then the complete response set $C(h, 1)$ is the convex hull of the point-source response set for the h-moment $F(h)$. That is,

$$C(h, 1) = \text{ch}\big(F(h)\big) \qquad .$$

PROOF Since X does not contain the coordinate origin, we see that $C(h, 1)$ may be expressed

$$C(h, 1) = \left\{ c : c = \int x[h] r(x) \frac{f(x)}{x[h]}\, dx \quad \text{for all} \quad r \in R(h, 1) \right\}$$

Since $r(x)$ in this equation belongs to $R(h, 1)$, it is evident that $x[h]r(x)$ is a normalized probability density function on X. Consequently $C(h, 1)$ belongs to any convex set containing $F(h)$. Thus $C(h, 1)$ belongs to the convex hull of $F(h)$. Conversely, since the set $R(h, 1)$ is unconstrained, it contains elements $r(x)$ such that $x[h]r(x)$ is an atomic density concentrated at any given point in X. In other words, for any x' in X, $R(h, 1)$ contains the following function

$$r(x) = \frac{\delta(x - x')}{x[h]}$$

where δ is the Dirac delta function [4]. From the integral properties of the delta function it results that $C(h, 1)$ contains $F(h)$. Since $C(h, 1)$ belongs to the convex hull of $F(h)$, $C(h, 1)$ must in fact equal this convex hull. QED

Let us suppose that we wish to design an assay system for determining the h-moment of the spatial distribution of the analyte. The measure-of-performance must indicate what values of the h-moment we are able to distinguish. The measure we shall develop is the *relative resolution* for the h-moment, denoted $z(h)$, and defined as follows. Any spatial distribution of analyte whose h-moment equals u is distinguishable from any spatial distribution of analyte whose h-moment is $v > u$, if and only if $v > z(h)u$. Thus $z(h)$ is a direct generalization of the relative mass resolution (without statistical uncertainty) z defined in Chapter 2. In fact, $z = z(0)$. The major result of our analysis will be an algorithm for evaluating $z(h)$. For convenience we re-iterate the following definition given in Chapter 2.

DEFINITION 2 Let A be a subset of the real n-dimensional Euclidean space $E(n)$. The *expansion* of A, if it exists, is denoted $e(A)$ and is defined by

$$e(A) = \sup\{y : A \cap yA \neq \emptyset\}$$

When there are elements f and g of A such that

$$g = e(A)f$$

then A is said to be *self-expanded*.

The following Lemma establishes that an important class of sets are self expanded. The proof is given elsewhere [5].

LEMMA 1 Let A be a non-empty compact [6] subset of $E(n)$. If the expansion of A exists, then A is self-expanded.

The following Theorem illustrates the utility of the concept of expansion.

THEOREM 2 Let A be a compact convex subset of $E(n)$. For any positive numbers u and $v > u$,

$$uA \cap vA = \emptyset \qquad (3.2.8)$$

if and only if

$$\frac{v}{u} > e(A) \qquad (3.2.9)$$

PROOF Clearly eq.(8) is equivalent to

$$A \cap \frac{v}{u} A = \emptyset \qquad (3.2.10)$$

Thus eq.(9) implies eq.(10) by definition 2. Let us now assume that eq.(8) is true. Since A is convex, it is not difficult to show that the expansion of A is finite. Furthermore, since A is compact, it is self expanded (Lemma 1), so there are elements g and f of A for which

$$g = e(A)f \quad .$$

If relation (9) is false, then $v/u \le e(A)$. Since $v > u$, this can occur only if $e(A) > 1$. Since A is convex, $yg + (1-y)f$ belongs to A for any $0 < y < 1$. Thus $(ye(A) + 1 - y)f$ belongs to A. Since $e(A) > 1$, choose y as

$$y = \frac{v - u}{u(e(A) - 1)}$$

Thus

$$(ye(A) + 1 - y)f = \frac{v}{u} f \in A$$

which contradicts eq.(10). Thus, if eq.(8) holds, ineq.(9) must also be true. QED

To understand the significance of this result, let us consider the class of complete response sets, $\{C(h, u), \ u > 0\}$. For $v > u$, any spatial distribution of analyte whose h-moment equals u is distinguishable from any spatial distribution of analyte whose h-moment equals v if and only if

$$C(h, u) \cap C(h, v) = \emptyset \qquad . \qquad (3.2.11)$$

We have noted that the class of complete response sets has the property of proportionality. Hence Theorem 2 implies that this intersection is null if and only if

$$\frac{v}{u} > \mathrm{e}\big(C(h, 1)\big) \qquad .$$

In other words, the expansion of $C(h, 1)$ is precisely the relative resolution for the h-moment. This is a generalization of the result reached in Chapter 2, that the expansion of C (which equals $C(0, 1)$) is the relative mass resolution (which is the relative resolution for the zeroth-moment).

The expansion of $C(h, 1)$ may be evaluated by the method justified in section 2.9 for evaluating the expansion of the complete response set C. We re-iterate this algorithm in the following Theorem.

THEOREM 3 Let B be a compact subset of $E(n)$ and let $A = \mathrm{ch}(B)$. Let W be the set of real n-tuples w for which

$$\langle w, f \rangle \langle w, g \rangle > 0 \qquad (3.2.12)$$

for all f and g in B. The expansion of A, if it exists, is given by

$$\mathrm{e}(A) = \underset{w \in W}{\mathrm{minimum}} \ \underset{f, g \in B}{\mathrm{maximum}} \ \frac{\langle w, f \rangle}{\langle w, g \rangle} \qquad (3.2.13)$$

Since $C(h, 1)$ is the convex hull of $F(h)$, we may obtain the expansion of $C(h, 1)$ by replacing A by $C(h, 1)$ and B by $F(h)$ in Theorem 3.

We may summarize our results as follows. Suppose the point-source response set for the h-moment, $F(h)$, is compact, and that the expansion of the complete response set, $C(h, 1)$, exists. Then the relative resolution for the h-moment is equal to the expansion of $C(h, 1)$, which may be evaluated from $F(h)$ by the min-max algorithm of Theorem 3. This result is based on the fact that $C(h, 1)$ is the convex hull of $F(h)$ and that the class of complete response sets, $\{C(h, u), \ u > 0\}$, has the property of proportionality.

It will usually be easy to establish that $F(h)$ is compact, as follows. $F(h)$ is the image of X (the domain of the sample container), under the mapping $x \mapsto f(x)/x[h]$ where $f(x)$ is the point-source response function. If X is compact and if $f(x)/x[h]$ is continuous on X, then $F(h)$ is the continuous image of a compact set, and thus compact.

3.3 CONSTRAINED SPATIAL DISTRIBUTIONS

3.3.1 Evaluation of the Relative Resolution

In Chapter 1 we identified many assay applications in which the analyte material is randomly distributed in space. However it may be that not every conceivable distribution is in fact allowed or realized in practice. The range of allowed spatial distributions may be limited or constrained. For any particular application, let us denote by $\widetilde{R}(h, u)$ the set of allowed spatial distributions, integrable on X, whose h-moment equals u. This is a subset of the set $R(h, u)$ of all functions integrable on X whose h-moment equals u.

In the previous section we found that evaluation of the resolution capability of an assay system designed for unconstrained spatial distributions is based on the convexity and proportionality of the complete response sets, $C(h, u)$. By imposing constraints on the set of allowed spatial distributions we may eliminate either or both of these properties of the complete response sets. In this section we shall extend the algorithm for evaluating the relative resolution to the case where the response sets are convex, though not necessarily proportional.

The complete response set for allowed spatial distributions whose h-moment equals u is denoted $\widetilde{C}(h, u)$. This set is defined analogously to eq.(3.2.6) as

$$\widetilde{C}(h, u) = \left\{ c : c = \int r(x) f(x)\, dx \quad \text{for all} \quad r \in \widetilde{R}(h, u) \right\}$$
$$(3.3.1)$$

The following Theorem provides the basic tool for evaluating the resolution of the assay system.

THEOREM 4 If the response sets $\widetilde{C}(h, u)$ and $\widetilde{C}(h, v)$ are convex and compact, then

$$\widetilde{C}(h, u) \cap \widetilde{C}(h, v) = \emptyset \qquad (3.3.2)$$

if and only if there exists a real n-tuple, w, such that either

$$\underset{c \in \widetilde{C}(h,u)}{\text{maximum}} \langle w, c \rangle < \underset{c \in \widetilde{C}(h,v)}{\text{minimum}} \langle w, c \rangle \qquad (3.3.3)$$

or

$$\underset{c \in \widetilde{C}(h,v)}{\text{maximum}} \langle w, c \rangle < \underset{c \in \widetilde{C}(h,u)}{\text{minimum}} \langle w, c \rangle \qquad (3.3.4)$$

In other words, eq.(2) holds if and only if there is a real linear functional, defined by w, which maps $\widetilde{C}(h, u)$ and $\widetilde{C}(h, v)$ into disjoint intervals on the real line.

PROOF If either of relations (3) or (4) hold, there is a hyperplane, defined by the n-tuple w, which separates $\widetilde{C}(h, u)$ and $\widetilde{C}(h, v)$. Hence these response sets are disjoint and eq.(2) is true. Conversely, if eq.(2) holds then, since the response sets are convex and compact, there is a hyperplane separating them [7] , and thus either ineq.(3) or ineq.(4) must hold. QED

Since Theorem 4 does not assume proportionality of the response sets, it lacks the simplicity of Theorems 2 and 3. The distinctive feature of proportional convex sets is that, for $v > u$,

$C(h, v)$ is invariably above (and $C(h, u)$ below) the plane separating these response sets. For non-proportional sets this ordering need not hold, as we shall show by example in section 3.4. Furthermore, by imposing the condition of proportionality on the response sets for constrained distributions, we obtain the ordering which is characteristic of unconstrained sets. This is shown explicitly in the next Theorem.

THEOREM 5 Let the response sets $\tilde{C}(h, u)$ and $\tilde{C}(h, v)$ be compact, convex and satisfy the following condition of proportionality:

$$\tilde{C}(h, v) = \frac{v}{u}\, \tilde{C}(h, u) \qquad .$$

If the expansion of $\tilde{C}(h, u)$ exists, then, for $v > u$,

$$\tilde{C}(h, u) \cap \tilde{C}(h, v) = \emptyset \tag{3.3.5}$$

if and only if

$$\frac{v}{u} > e\big(\tilde{C}(h, u)\big) \tag{3.3.6}$$

PROOF Suppose ineq.(6) holds. If relation (5) does not hold then there are elements a and b of $\tilde{C}(h, u)$ and $\tilde{C}(h, v)$ respectively such that $b = a$. By the condition of proportionality there is an element f in $\tilde{C}(h, u)$ such that

$$b = \frac{v}{u} f = a$$

This contradicts ineq.(6) by the definition of the expansion since f and a both belong to $\tilde{C}(h, u)$. Hence eq.(5) must hold. Now suppose that eq.(5) holds and let $z = e\big(\tilde{C}(h, u)\big)$. If ineq.(6) is false then $z \geq v/u$. Since $\tilde{C}(h, u)$ is compact, it is self expanded (Lemma 1). Thus there are elements g and h of $\tilde{C}(h, u)$ such that $h = zg$. By the proportionality of $\tilde{C}(h, u)$ and $\tilde{C}(h, v)$ one obtains

$$\frac{v}{u} h \in \tilde{C}(h, v) \qquad \text{and} \qquad \frac{v}{u} g \in \tilde{C}(h, v) \qquad .$$

Employing the relation $h = zg$, the latter inclusion becomes

$$\frac{v}{zu} h \in \widetilde{C}(h, v) \qquad .$$

$\widetilde{C}(h, v)$ is convex, so for any $0 < y < 1$,

$$\left(\frac{yv}{u} + \frac{(1-y)v}{zu} \right) h \in \widetilde{C}(h, v) \qquad .$$

Since $z \geq v/u > 1$, we may choose

$$y = \frac{uz - v}{(z - 1)v}$$

from which it results that h belongs to $\widetilde{C}(h, v)$. But since h also belongs to $\widetilde{C}(h, u)$, this contradicts eq.(5). Hence relation (6) must be true, and the proof is complete. QED

3.3.2 Fundamental Response Sets

The calculation of the relative resolution of unconstrained distributions was considerably simplified by the proportionality of the class $\{C(h, u), \ u > 0\}$ of complete response sets and by the simple relation between the complete response set $C(h, 1)$ and the point-source response set $F(h)$, as expressed by the convexity theorem (Theorem 1 in Chapter 2). This convexity relation allows the expansion of $C(h, u)$ to be calculated by a min-max search performed on $F(h)$, as given in Theorem 3. The situation we confront with constrained spatial distributions is more complex. In Theorem 4, the extrema are defined on the complete response sets $\widetilde{C}(h, u)$ and $\widetilde{C}(h, v)$. Because of the arbitrary nature of the constraints on the spatial distributions we are not always able to define a point-source response set whose convex hull equals the complete response set.

However, in some circumstances the following technique may be used to define an analog of the point-source response set. Recall that $\widetilde{R}(h, u)$ is the set of allowed spatial distributions with

h-moment equal to u. If $\tilde{R}(h, u)$ is compact and convex, it may be possible to express $\tilde{R}(h, u)$ as the convex hull of some set $D(h, u)$ of *fundamental distributions*. That is, we may be able to find a set $D(h, u)$ such that

$$\tilde{R}(h, u) = \mathrm{ch}\big(D(h, u)\big) \qquad (3.3.7)$$

Quite often the physical definition of the assay problem will lead to a natural choice of $\tilde{R}(h, u)$ and its fundamental set $D(h, u)$.

If $\tilde{R}(h, u)$ is a compact convex set for which a set of fundamental distributions exists, we are able to define an analog of the point-source response set. The *fundamental response set* $\tilde{F}(h, u)$ is defined as

$$\tilde{F}(h, u) = \left\{ g : g = \int d(x) f(x)\, dx \qquad \text{for all} \qquad d \in D(h, u) \right\}$$
$$(3.3.8)$$

Combining eqs.(1), (7) and (8) it is not difficult to show that the complete response set is the convex hull of this fundamental response set. That is

$$\tilde{C}(h, u) = \mathrm{ch}\left(\tilde{F}(h, u) \right) \qquad (3.3.9)$$

Now the inner product $\langle w, c \rangle$ in eqs.(3) and (4) is in fact a linear functional on the set $\tilde{R}(h, u)$ of allowed spatial distributions. That is, for any r and s in $\tilde{R}(h, u)$ and for any real numbers x and y

$$\langle w, c(xr + ys) \rangle = x \langle w, c(r) \rangle + y \langle w, c(s) \rangle \qquad (3.3.10)$$

A linear functional achieves the same extremal values on a compact set S and on any subset whose convex hull equals S [9]. Thus the extrema in Theorem 4 may be sought on $\tilde{F}(h, u)$ rather than on $\tilde{C}(h, u)$. Furthermore each element of $\tilde{F}(h, u)$ is specified by a fundamental distribution in $D(h, u)$. Thus the extrema may be sought on $D(h, u)$ as well. Explicitly,

$$\underset{c \in \widetilde{C}(h,u)}{\text{extremum}} \langle w, c \rangle = \underset{g \in \widetilde{F}(h,u)}{\text{extremum}} \langle w, c(g) \rangle \qquad (3.3.11)$$

$$= \underset{d \in D(h,u)}{\text{extremum}} \langle w, c(d) \rangle \qquad (3.3.12)$$

We shall find this formulation useful in a number of examples to follow.

3.3.3 Inclusion of Statistical Uncertainty

Theorem 4 provides a means of determining whether or not every allowed spatial distribution whose h-moment equals u is distinguishable from every allowed spatial distribution whose h-moment equals v, in the absence of statistical uncertainty of measurement. As we have stressed before, the design process is likely to comprise a search for the optimal compromise between the constraints of spatial and statistical uncertainty. It is thus essential to extend the algorithm of Theorem 4 to include the statistical uncertainty of measurement. This is readily done by applying the approach presented in section 2.11

Let $\sigma(\langle w, tc \rangle)$ represent the statistical standard deviation of the linear combination of measurements $\langle w, tc \rangle$. This standard deviation may be evaluated as in eq.(2.11.7). Thus the values u and v of the h-moment are distinguishable, to a specified level of statistical confidence, if there exists a real n-tuple w such that either

$$\underset{c \in \widetilde{C}(h,u)}{\text{maximum}} \Big(\langle w, tc \rangle + b\sigma\big(\langle w, tc \rangle\big) \Big)$$

$$< \underset{c \in \widetilde{C}(h,v)}{\text{minimum}} \Big(\langle w, tc \rangle - b\sigma\big(\langle w, tc \rangle\big) \Big)$$

or

$$\underset{c \in \widetilde{C}(h,v)}{\text{maximum}} \Big(\langle w, tc \rangle + b\sigma\big(\langle w, tc \rangle\big) \Big)$$

$$< \underset{c \in \widetilde{C}(h,u)}{\text{minimum}} \Big(\langle w, tc \rangle - b\sigma\big(\langle w, tc \rangle\big) \Big)$$

where b is a non-negative number indicating the level of statistical confidence required.

3.3.4 Convexity and Compactness of $\widetilde{C}(h, u)$

In order to apply Theorem 4 it is necessary to establish the convexity and compactness of the sets $\widetilde{C}(h, u)$. Lemmas 2 and 5 will treat this problem.

A simple (though not exhaustive) criterion can be established for determining the convexity of $\widetilde{C}(h, u)$, as shown by the next Lemma. Let $\widetilde{R}$ represent the set of all allowed spatial distributions of any moment.

LEMMA 2 If $\widetilde{R}$ is convex, then $\widetilde{R}(h, u)$ and $\widetilde{C}(h, u)$ are convex.

PROOF First let us note that $\widetilde{R}(h, u)$ (the set of all allowed spatial distributions whose h-moment equals u) is the intersection of $\widetilde{R}$ and of $R(h, u)$ (the set of all functions integrable on X whose h-moment equals u). That is,

$$\widetilde{R}(h, u) = \widetilde{R} \cap R(h, u)$$

Thus $\widetilde{R}(h, u)$ is the intersection of convex sets and thus convex. Now, for any two elements r and s of $\widetilde{R}(h, u)$, let g and h be elements of $\widetilde{C}(h, u)$ defined by

$$g = \int r(x) f(x) \, dx \qquad \text{and} \qquad h = \int s(x) f(x) \, dx$$

where f is the point-source response function. Then, for any $0 < y < 1$, $yg + (1-y)h$ also belongs to $\widetilde{C}(h, u)$ because $yr + (1-y)s$ belongs to $\widetilde{R}(h, u)$. Thus $\widetilde{C}(h, u)$ is convex. QED

We shall now develop a criterion for determining the compactness of $\widetilde{C}(h, u)$. We first require the following standard definitions and two preliminary Lemmas.

DEFINITION 3 Let Q be a class of real scalar functions defined on a subset Y of a metric space with metric d. The class Q is *uniformly bounded* if there exists a positive number t such that

$|q(y)| < t$ for all y in Y and all q in Q. The family Q is *equicontinuous* on Y if, given any $a > 0$ there exists a number $b > 0$ such that $d(y, z) < b$ implies that $|g(y) - q(z)| < a$ for all y and z in Y and all q in Q. A subset of a metric space is *relatively compact* if its closure is compact.

Let $K(Y)$ represent the set of all real scalar continuous functions on a subset Y of a Euclidean metric space. Thus for instance $K(X)$ is the set of all continuous functions on the domain of the sample container. $K(Y)$ is a metric space with the following *uniform* metric.

$$d_u(r, s) = \underset{t \in Y}{\text{maximum}} \, |r(t) - s(t)| \qquad .$$

The usual Euclidean metric is

$$d_e(x, y) = \left(\sum (x_i - y_i)^2 \right)^{1/2}$$

The following Lemma is well-known [8] and will be useful.

LEMMA 3   Let (Y, d_u) be a compact metric space where d_u is the uniform metric. A class $G(Y)$ of continuous functions on Y is relatively compact in the set $K(Y)$ of all continuous functions on Y if and only if $G(Y)$ is uniformly bounded and equicontinuous.

We need one more preliminary Lemma before presenting a criterion for the compactness of $\widetilde{C}(h, u)$.

LEMMA 4   Let $g(x)$ be an n-dimensional real function defined and bounded on a bounded subset Y of a Euclidean space. Let V be the volume of the region Y. That is $V = \int dy$. Define the mapping, h, from $K(Y)$ into $E(n)$ as

$$h(k) = \int_Y k(y) g(y) \, dy$$

The mapping h is continuous at each point k in $K(Y)$.

PROOF   We must show that, for all $a > 0$, there is a number $b > 0$ such that

$$d_u(k, k') < b \qquad \text{for} \qquad k, k' \in K(Y) \qquad (3.3.13)$$

implies that
$$d_e\big(h(k), h(k')\big) < a \qquad . \qquad (3.3.14)$$

The Lemma is trivial if g vanishes everywhere on Y, so suppose this is not the case. Since g is bounded on Y, there is a number $\widehat{g}$ such that
$$\widehat{g} = \underset{1 \le i \le n}{\text{maximum}} \ \underset{y \in Y}{\text{maximum}} \ |g_i(y)| \qquad .$$

For any $a > 0$, choose
$$b = \frac{a}{2\widehat{g}V\sqrt{n}}$$

Then, ineq.(13) implies that
$$\underset{y \in Y}{\text{maximum}} \ |k(y) - k'(y)| < b$$

Also, by the definition of the Euclidean metric,
$$d_e\big(h(k), h(k')\big)$$

$$= \left(\sum_i \left(\int_Y k(y)g_i(y)\,dy - \int_Y k'(y)g_i(y)\,dy \right)^2 \right)^{1/2}$$

$$= \left(\sum_i \left(\int_Y (k(y) - k'(y))g_i(y)\,dy \right)^2 \right)^{1/2}$$

By the Cauchy-Schwarz inequality this becomes

$$d_e\big(h(k), h(k')\big) \le \left(\sum_i \int (k(y) - k'(y))^2\,dy \int g_i^2(y)\,dy \right)^{1/2}$$

$$\le \widehat{g}\left(V \sum_i \int (k(y) - k'(y))^2\,dy \right)^{1/2}$$

$$\le \widehat{g}\left(V \sum_i \int b^2\,dy \right)^{1/2}$$

$$= \widehat{g}Vb\sqrt{n} = \frac{a}{2} \qquad .$$

which establishes ineq.(14) and completes the proof. QED

LEMMA 5 If the response function f is bounded on X and if X is compact and if $\tilde{R}$ is a closed, uniformly bounded and equicontinuous subset of $K(X)$, then $\tilde{C}(h, u)$ is compact.

PROOF Our proof will be in three parts. In part (a) we shall show that $\tilde{R}$ and the set $\tilde{C}$ of all possible responses are compact sets, in part (b) that $\tilde{R}(h, u)$ is compact and in part (c) that $\tilde{C}(h, u)$ is compact.

(a) (X, d_e) is a compact metric space where d_e is the Euclidean metric defined earlier. Since $\tilde{R}$ is a subset of the set $K(X)$ of all continuous functions on X, and since $\tilde{R}$ is uniformly bounded and equicontinuous, we conclude from Lemma 3 that $\tilde{R}$ is relatively compact in $K(X)$. Since $\tilde{R}$ is closed, it is compact. The set of all possible responses is defined as

$$\tilde{C} = \left\{ c : c = \int r(x) f(x) \, dx \qquad \text{for all} \qquad r \in \tilde{R} \right\}$$

By Lemma 4 and the definition of $\tilde{C}$, we see that $\tilde{C}$ is a continuous image of the compact set $\tilde{R}$. Hence $\tilde{C}$ is compact.

(b) Let h be a vector of non-negative integers, and define the mapping $T : K(X) \mapsto E(1)$ by

$$T(r) = \int x[h] r(x) \, dx \qquad \text{for} \qquad r \in K(X).$$

Since X is bounded, $x[h]$ is bounded on X. Thus, by Lemma 4, T is a continuous mapping. For a real number u, the pre-image in $K(X)$ of the set $\{u\}$ is precisely $R(h, u)$. Since $\{u\}$ is a closed set and T is a continuous mapping, $R(h, u)$ is a closed set. Now, the set of allowed functions whose h-moment equals u is given by

$$\tilde{R}(h, u) = R(h, u) \cap \tilde{R}$$

which is the intersection of closed sets and thus closed. Since $\tilde{R}(h, u)$ is the closed subset of the compact set $\tilde{R}$, we conclude that $\widetilde{R}(h, u)$ is compact.

(c) Recall the definition of $\widetilde{C}(h, u)$ in eq.(1). Since f is bounded on X, we conclude from Lemma 4 that f maps $\widetilde{R}(h, u)$ continuously into $\widetilde{C}(h, u)$. Since $\widetilde{R}(h, u)$ is compact, we conclude that $\widetilde{C}(h, u)$ is also compact. QED

3.4 EXAMPLE: SIMPLE CONSTRAINED DISTRIBUTIONS

In this section we shall examine a very simple moment-assay of constrained spatial distributions. Let X be the interval $[0, 1]$ on the real line. The set $\widetilde{R}$ of all allowed spatial distributions is the convex hull of the set of two *fundamental distributions*. These fundamental distributions are

$$r_1(x) = -x + 1 \qquad , \qquad r_2(x) = 2x \qquad .$$

Elements of $\widetilde{R}$ are averages of the two fundamental distributions, and may be denoted

$$r(x, t) = tr_1(x) + (1 - t)r_2(x)$$

for $0 \le t \le 1$. The zero and first moments of the spatial distribution $r(x, t)$ are

$$m(0) = 1 - \frac{t}{2}$$

$$m(1) = \frac{2}{3} - \frac{t}{2} \qquad .$$

From this we see that each value of $m(0)$ or $m(1)$ is attained by at most one element of $\widetilde{R}$. Thus the sets of allowed distributions whose zeroth or first moment equals u are, respectively,

$$\widetilde{R}(0, u) = \{r(x, 2 - 2u)\}$$

$$\widetilde{R}(1, u) = \left\{r\left(x, \frac{4}{3} - 2u\right)\right\}$$

Let us suppose that the point-source response function is

$$f(x) = (1 - x)^2$$

Hence the measured response to the spatial distribution $r(x, t)$ will be

$$c(r) = \int r(x, t) f(x) \, dx = \frac{t + 2}{12} \qquad .$$

Comparing this relation with the expressions for $m(0)$ and $m(1)$ reveals that the response decreases monotonically with increasing value of either moment. This is manifested in the complete response sets, which are

$$\tilde{C}(0, u) = \left\{ \frac{2 - u}{6} \right\} \qquad , \qquad 1/2 \leq u \leq 1$$

$$\tilde{C}(1, u) = \left\{ \frac{5 - 3u}{18} \right\} \qquad , \qquad 1/6 \leq u \leq 2/3$$

In other words, for $u < v$, the response set $\tilde{C}(0, u)$ is "above" the response set $\tilde{C}(0, v)$, unlike the proportional ordering which is characteristic of unconstrained spatial distributions. Nevertheless, the resolution capability of this very simple assay problem is quite good: any value of either moment is unambiguously distinguishable.

It should be noted that the reversed ordering of the complete response sets which we saw in this example is not universal. For instance, let us replace the quadratic response function with the following one:

$$f(x) = (1 - x)^{1/2}$$

Then the response to the spatial distribution $r(x, t)$ is

$$c(r) = \frac{8 - 2t}{15}$$

and the complete response sets are

$$\widetilde{C}(0, u) = \left\{ \frac{4 + 4u}{15} \right\}$$

$$\widetilde{C}(1, u) = \left\{ \frac{16 + 2u}{45} \right\}$$

Thus the complete response sets are ordered even though the class of response sets does not have the property of proportionality.

3.5 EXAMPLE: CONSTRAINED NORMAL DISTRIBUTIONS

In this section we shall consider a more complex example, and illustrate the application of our results to the design of the assay system. Let the domain of the sample, X, be a large interval centered at the origin of the real line. Since all the functions involved will tend to zero for large arguments, we may consider X as covering the entire real line. The set of fundamental distributions comprises four normal distributions,

$$n(x, i, j) = \frac{1}{\sqrt{2\pi v_j}} \exp\left(-\frac{(x - m_i)^2}{2v_j} \right) \quad , \quad i, j = 1 \text{ or } 2$$

As before, the set of allowed spatial distributions, $\widetilde{R}$, is the convex hull of the set of fundamental distributions.

We are interested in assaying the first moment of the spatial distribution, which may assume any of the values

$$m(t) = tm_1 + (1 - t)m_2 \quad , \quad 0 \le t \le 1 \quad .$$

Those convex combinations of two fundamental distributions whose first moments equal $m(t)$ are

$$q(t, i, j) = tn(x, 1, i) + (1 - t)n(x, 2, j) \quad , \quad i, j = 1 \text{ or } 2 \quad .$$

Let us denote the set of these four distributions by $D\big(1, m(t)\big)$. These four distributions are the fundamental distributions of the set $\widetilde{R}(1, m(t))$ of allowed spatial distributions whose first moment equals $m(t)$. That is,

$$\widetilde{R}(1, m(t)) = \mathrm{ch}(D(1, m(t))$$

Let the response function be

$$f(x) = \exp\left(-w(x - x')^2\right)$$

where w is a known positive constant and x' is a known fixed point on the real line. Thus the sensitivity of the detector decreases as the source material is displaced from the point x'. We shall suppose that the point x' is related to the physical position of the detector. The aim of our design-analysis is to choose the best value for x'. The final choice of x' will depend on the range of values of $m(t)$ which we wish to measure. Our design-analysis will proceed as follows. For a given value of x' we shall determine the range of values of the first moment which are distinguishable. Different values of x' will generate sensitivity to different ranges of values of $m(t)$. In this way the value of x' may be suited to the needs of a particular assay.

Theorem 4 constitutes the basis of our determination of the resolution capability. Since the assay involves only a single measurement, the complete response set $\widetilde{C}(1, m(t))$ is a closed interval on the real line. To determine the resolution capability we must find the end points of this interval. Since $\widetilde{R}(1, m(t))$ is a compact set and is the convex hull of $D(1, m(t))$, we must find

$$c_{\max}(t) = \underset{q \in D(1, m(t))}{\text{maximum}} \int q(x) f(x)\, dx$$

$$c_{\min}(t) = \underset{q \in D(1, m(t))}{\text{minimum}} \int q(x) f(x)\, dx$$

Let us define

$$h(i, j) = \int n(x, i, j) f(x)\, dx$$

$$= (2wv_j + 1)^{-1/2} \exp\left(\frac{-w}{2wv_j + 1}(m_i - x')^2\right)$$

Then

$$\int q(t,i,j)f(x)\,dx = t\,h(1,i) + (1-t)h(2,j)$$

and

$$c_{\max}(t) = tH(1) + (1-t)H(2)$$

and

$$c_{\min}(t) = t\,h(1) + (1-t)h(2)$$

where

$$H(i) = \underset{j=1,2}{\text{maximum}}\, h(i,j)$$

$$h(i) = \underset{j=1,2}{\text{minimum}}\, h(i,j)$$

From Theorem 4 we conclude that any spatial distribution whose first moment equals $m(t)$ is distinguishable from any spatial distribution whose first moment equals $m(t')$ if and only if either

$$c_{\max}(t) < c_{\min}(t')$$

or

$$c_{\max}(t') < c_{\min}(t)$$

It is instructive to consider a numerical example. Let $m_1 = 0$, $m_2 = 1$, $v_1 = 1/4$, $v_2 = 1$ and $w = 1$. In figure 3.5.1 we display the distinguishable values of $m(t)$ and $m(t')$, for various choices of x'. In each frame we plot t' versus t. The regions of distinguishable values of the first moment are shaded. The value of x' is shown in the center of each frame. Figure 3.5.1(a) displays the resolution capability for positive values of x'. We see that the first-moment resolution capability decreases rapidly as x' rises above 0. However, in each of these six frames we find that values of $m(t)$ are indistinguishable in the vicinity of $m(t) = 0$ ($t = 1$; the upper right corner of each frame). This indicates that the detector is unable to resolve first-moment values near the detector position x'. These results may confirm the designer's intuitive feeling, and they provide a rigorous quantitative evaluation of the resolution capability. In figures 3.5.1(b) and 3.5.1(c) the detector

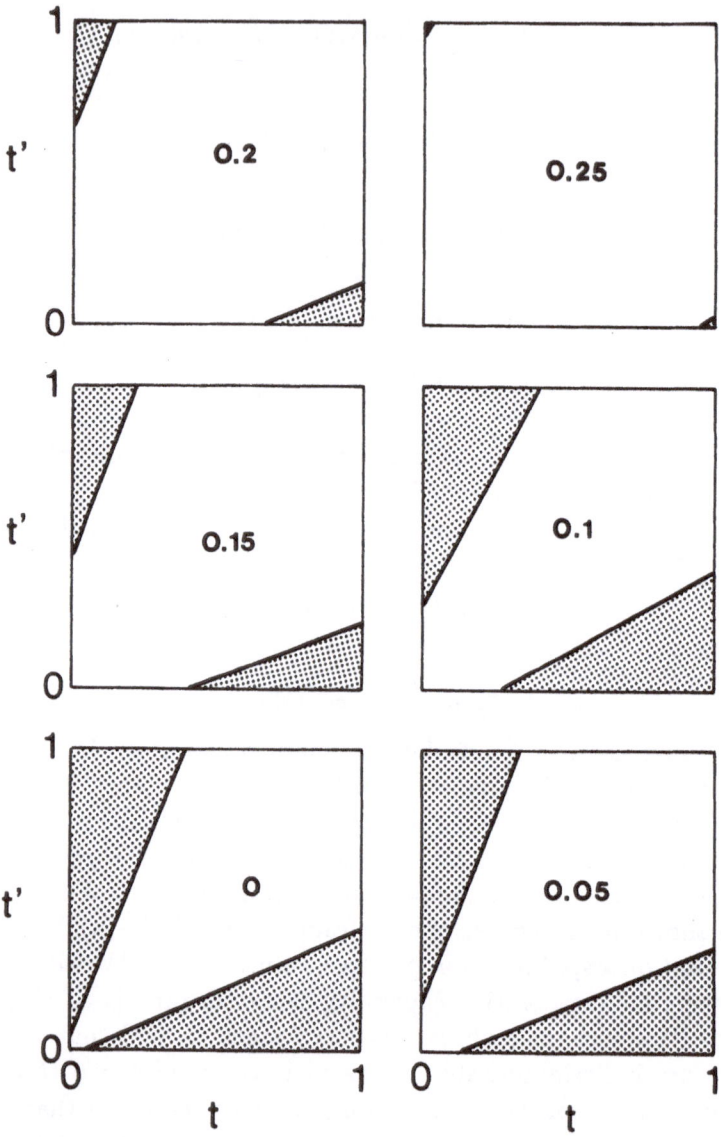

Figure 3.5.1 (a) Regions of resolvable first moment for positive detector positions. The value of x' is shown in each frame.

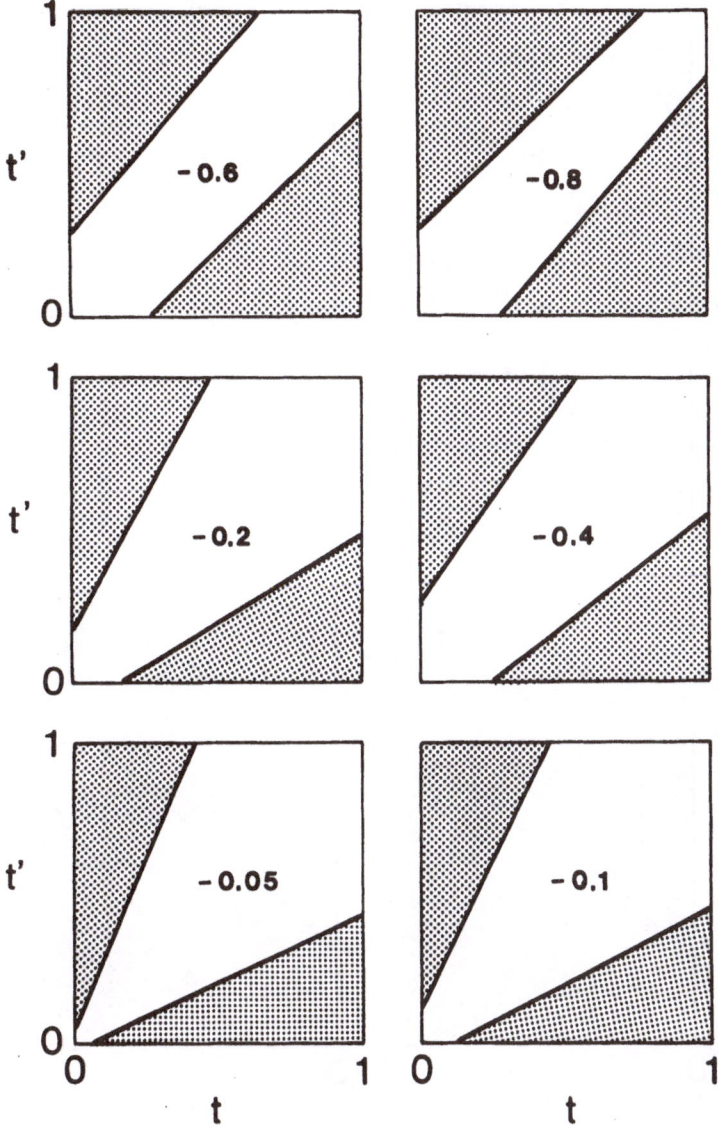

Figure 3.5.1 (b) Regions of resolvable first moment for negative detector positions. The value of x' is shown in each frame.

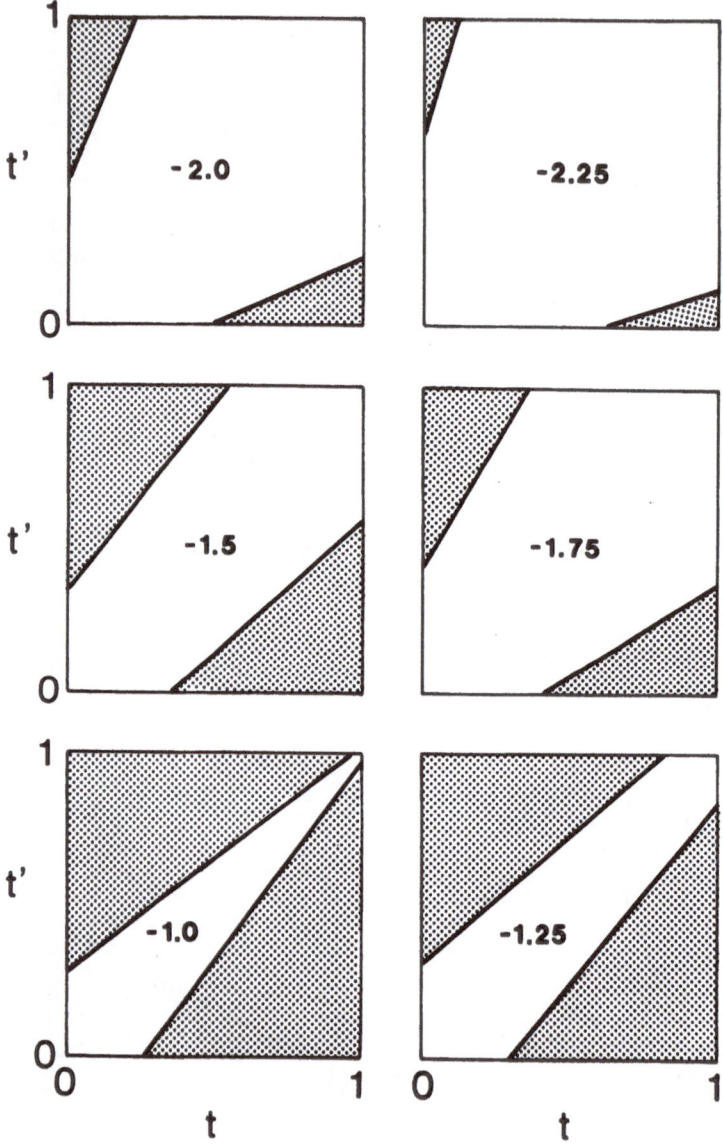

Figure 3.5.1 (c) Regions of resolvable first moment for negative detector positions. The value of x' is shown in each frame.

is moved to progressively more negative values. The resolution steadily improves, until about $x' = -1$, after which the resolution gradually deteriorates as the detector is removed further from the domain of the sample. It is interesting to compare the cases $x' = 0$ and $x' = -1$. The region of resolvable first-moment values is more extensive in the latter case. However, if the values of interest are close to $m(t) = 1$ ($t = 0$; the lower left corner of the frame), the former case may be preferred.

3.6 EXAMPLE: METEOROLOGICAL MEASUREMENTS

A challenging assay problem arises in the measurement of the transport and dispersal in the atmosphere of local releases of gases. Such measurements occur frequently in air pollution studies. The complexity of flow patterns in atmospheric boundary layers may generate quite variable and heterogeneous spatial distributions of the assayed material. In one study of nocturnal drainage wind flows in a complex terrain [10], air samplers were clustered several score meters from each other in areas 1 to 6 kilometers from the gas-release point. Comparison of these samplers showed consistent differences by a factor of 10 or more from one sampler to another, even when averaged over several hours. Evidently, wind flow patterns which have developed over a region of many square kilometers generate strong sustained local variations in gas concentration.

This variation of gas concentration (and hence sampler response) presents a very serious problem of calibration. The designer of a network of air samplers must decide how to distribute the measurement stations so as to be able to overcome the unknown local variation of gas concentration. For instance, by placing many samplers very close together (in the above case, probably every few meters) one is able to map out the local fluctuations and thus determine the overall average concentration. This however calls for an extraordinary number of samplers, which is probably impractical in most circumstances.

The aim of the designer must be to find efficient sampler-

configurations which are insensitive to the local heterogeneity of
the gas concentration. Since the possible deployments of samplers are innumerable, a systematic approach based on a physically
meaningful and computationally efficient evaluation is essential.
The concept of relative resolution and the min-max algorithm are
well suited to this task. The design analysis is two-fold. First, it is
necessary to establish as good an estimate as possible of the range
of possible local concentration profiles in each region in which a
cluster of samples may be located. That is, one must make some
approximation to the set $\widetilde{R}(0,1)$ of allowed spatial distributions of
1 unit of assayed gas. Since the gas concentration is very low and
does not influence the wind flow patterns, the set of allowed spatial distributions of u units of gas is just u times the set of allowed
spatial distributions of one unit of gas. That is, the class of sets
$\{\widetilde{R}(0,u),\ u>0\}$ has the property of proportionality. Each sampler measures the concentration only in its immediate surround-
ings. Thus for a sampler at position x', its point-source response
function may be represented by the Dirac delta, $\delta(x-x')$. Now
suppose that for a given region of interest we contemplate employing n detectors at points $x_1,\ldots,x_n$. The complete response set
is

$$\widetilde{C}(0,1) = \left\{ c : c = \big(r(x_1),\ldots,r(x_n)\big) \qquad \text{for all} \qquad r \in \widetilde{R}(0,1) \right\}$$
$$(3.6.1)$$

If $\widetilde{R}(0,1)$ is the convex hull of a fundamental set $D(0,1)$ of allowed spatial distributions, then $\widetilde{C}(0,1)$ is the convex hull of a
fundamental response set $\widetilde{F}(0,1)$ defined by

$$\widetilde{F}(0,1) = \left\{ g : g = \big(r(x_1),\ldots,r(x_n)\big) \qquad \text{for all} \qquad r \in D(0,1) \right\}$$
$$(3.6.2)$$

The second stage of the analysis is the evaluation of the expansion of $\widetilde{C}(0,1)$ for various numbers and placements of the samplers.
This may be readily performed by computer calculation with the
aid of the min-max algorithm of Theorem 3.

We shall illustrate the design analysis for this type of problem with a simple example. Let us consider, as the area in which the samplers are to be located, a small region in which the concentration may be expected to increase toward the middle of the region, but there is considerable uncertainty regarding the position of the peak concentration. Such a situation may arise, for instance, if the sampled region is the cross section of a small canyon. Let the domain of the region be $X = [0, 1]$, and let the fundamental set of allowed spatial distributions be

$$D(0, 1) = \left\{ r : r(x) = 2\sin^2(\pi x + b) \quad , \quad b \in B \right\} \qquad (3.6.3)$$

where B is the set of real numbers on a known interval $[b_1, b_2]$. In figure 3.6.1 are shown some examples of these fundamental distributions. The values of b for curves 1, 2 and 3 are 0, 1/8 and 1/2, respectively. The relative mass resolution for n detectors at points $x_1, \ldots x_n$ is given by

$$z(0) = \operatorname*{minimum}_{w \in W} \; \operatorname*{maximum}_{a, b \in B} \; \frac{\sum w_i \sin^2(\pi x_i + a)}{\sum w_i \sin^2(\pi x_i + b)} \qquad (3.6.4)$$

where W is the set of real n-tuples for which

$$\langle w, c \rangle \langle w, d \rangle > 0 \qquad (3.6.5)$$

for all c and d in $\widetilde{F}(0, 1)$.

In figure 3.6.2 we show the relative mass resolution for a single detector versus the detector location. Plots for two different choices of the set B are shown. In curve 1, the phase b varies between $-1/8$ and $1/8$, while in curve 2 the phase varies from $-1/2$ to $1/2$. In both cases the best detector position is at the centroid of the range of variation of the allowed spatial distributions. For small variation of the allowed phase shift of the spatial distribution (curve 1) the best resolution is quite good ($z(0) = 1.016$ at $x = 0.5$). However the resolution is somewhat sensitive to uncertainty in locating the centroid of the spatial distributions. That is,

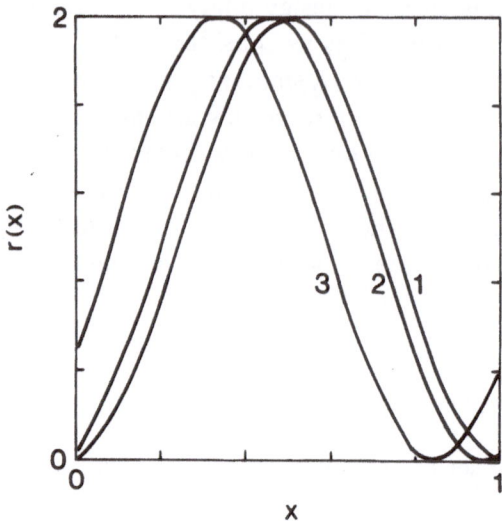

Figure 3.6.1 Fundamental spatial distributions.

if the spatial distributions vary by a phase shift of $\pm 1/8$ around the point $x = 0.4$, rather than around $x = 0.5$, the relative resolution will be 1.178 if the detector is located at $x = 0.5$. For large allowed phase shift the best relative resolution is much poorer ($z(0) = 1.30$ at $x = 0.5$), and the resolution deteriorates rapidly if the detector is not located at the centroid of $\widetilde{R}(0, 1)$.

The one-detector resolution capability should probably be considered inadequate because of the sensitivity of $z(0)$ to imprecise knowledge of the centroid of the spatial distributions. It is thus necessary to examine multiple detector configurations. Even for 2 detectors the range of possibilities is large, and an automated computerized evaluation of $z(0)$ for a large selection of configurations is appealing. One finds that as the detectors are spaced progressively further apart the resolution steadily improves, and the sensitivity of the measurement to properly locating the centroid decreases. Indeed, when the distance between the detectors equals $1/2$, the relative mass resolution is unity, and is entirely insensitive to the location of the mid-point between the detectors.

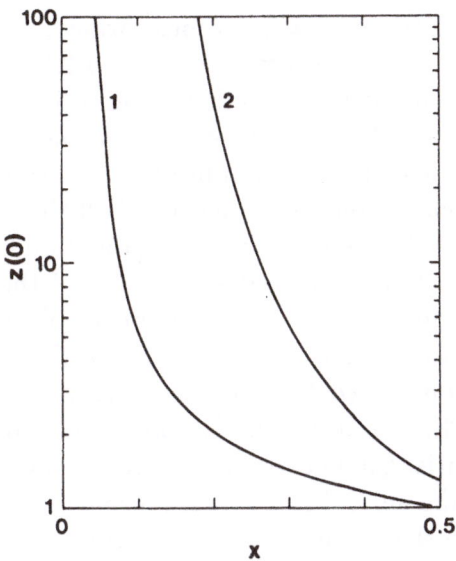

Figure 3.6.2   The relative mass resolution for a single detector versus the detector position.

This may be seen analytically by using elementary trigonometric identities to show that

$$\sin^2\left(\pi x_1 + b\right) + \sin^2\left(\pi(x_1 + 1/2) + b\right) = 1 \qquad (3.6.6)$$

for any values of x_1 and b. Thus the numerator and denominator of the right hand side of eq.(3.6.4) are equal (for $w_1 = w_2 = 1$) and so $z(0) = 1$.

3.7 EXAMPLE: ASSAY OF A PULMONARY AEROSOL

3.7.1 Formulation

Monitoring the lung burden of plutonium aerosol particles in radiation workers in the nuclear industry is an important and difficult task. It is necessary for health-safety reasons to detect very small quantities of plutonium. It is usually desirable to be able to detect 2 to 4 nCi.

The assay is based on passively measuring the L X-rays of plutonium, which range in energy from 13 to 20 keV [11]. The absorption of this low energy radiation is quite significant in lung tissue, soft tissue and bone, as expressed by the half-thickness: the distance in which the radiation-intensity is reduced by 50%. For lung tissue, soft tissue and bone the half-thicknesses at 17 keV are 22, 7.2 and 1.4 mm, respectively [12]. In addition to the strong absorption of the fluorescent radiation of the plutonium, the fluorescent count rate is typically about one fifth of the gross count rate. Due to this low signal to noise ratio, the phoswich detector is commonly used for this measurement because of its high efficiency and good background-suppression capability [13].

While the statistical uncertainty in the assay of the lung-burden of plutonium aerosol is quite significant, the spatial uncertainty arising from the unknown spatial distribution of the aerosol is also an important factor. In general, the ability to predict the spatial distribution of the aerosol is very limited. The spatial distribution is governed by a large number of mechanisms whose properties and interrelations are inadequately understood. Much effort has been devoted to the study of the distribution of aerosol particles between the major regions of the respiratory tract. Typically three regions are identified: the nasopharyngeal region, the tracheobronchial region and the pulmonary region. The major factors governing the aerosol distribution among these regions are the aerosol size and weight, the intra-lung geometry, the nature of the breathing cycle and the details of the air flow patterns. Also of importance is the duration of time since inhalation of the aerosol, during which time lung clearance mechanisms may operate [14]. Less is known about how the aerosol is distributed within each of the three major regions of the respiratory tract [15].

A final factor complicating the assay of the plutonium lung burden is the variable absorption characteristics of the lung tissue and of the chest wall. The lung and chest wall form a highly heterogeneous matrix in which the aerosol is distributed. The details of the spatial variation of the absorption characteristics of this matrix influence the measured activity due to the strong absorption of the L X-rays of plutonium.

In this section we shall apply the convexity theorem and the concept of relative mass resolution to the design of a system for assaying the lung burden of a plutonium aerosol. We shall limit our study to an analysis of the spatial uncertainty. The design question which we shall address is whether measurement from one side of the patient is sufficient or whether measurements from both posterior and anterior surfaces are warranted.

Such an analysis can be no better than the models on which it is based. Models are needed for describing the range of allowed spatial distributions of the aerosol in the respiratory tract, and for describing the radiation transport in the lung tissue and chest wall. We have stressed that there is considerable uncertainty in the construction of both of these models. No attempt will be made here to develop realistic models. Simplistic models will be employed to illustrate the nature of the analysis and to demonstrate the type of information which may be attained. It will be evident that more realistic models may be readily substituted.

The respiratory tract is a complex three dimensional structure. The response of a phoswich detector located at the patient's chest varies as a point source is moved throughout the spatial domain of the "sample". However the most important spatial degree of freedom is the depth of the source on the anterior-posterior axis. The response of a detector on the anterior surface increases by a factor of 200 or more as the point source moves from the back to the front of the lung, and increases by a factor of two as the point source moves from the midplane to the upper third of the lung. Furthermore, the variation of the count rate in response to variation in the depth of the source is fairly accurately represented by an exponential function.

In light of these data, we shall adopt a one-dimensional model of the lung. The spatial domain X of the sample is the unit interval $[0, 1]$ which represents the anterior-posterior axis of the patient. Furthermore the response is assumed to show front-back symmetry. Thus the point-source response function of a detector on the anterior surface of the patient is

$$f_1(x) = K\,e^{-bx} \qquad\qquad (3.7.1)$$

while the point-source response function of a posterior detector is

$$f_2(x) = K\,e^{-b(1-x)} \qquad\qquad (3.7.2)$$

The constant b is chosen to equal 5.30, which yields a decrease in the response by a factor of 200 across the lung.

3.7.2 Unconstrained Spatial Distributions

The design analysis is particularly simple if the plutonium may adopt any conceivable spatial distribution. While this is certainly not a realistic model of the range of allowed spatial distributions of the aerosol, this analysis serves as a preliminary assessment of the problem and provides an upper bound on the spatial uncertainty.

For a single detector, the complete response set is simply the closed interval from the minimum to the maximum responses of the detector:

$$C = [f_1(1), f_1(0)] = [K\,e^{-b}, K] \qquad\qquad (3.7.3)$$

Thus the relative mass resolution is

$$z = e^b = 200 \qquad . \qquad\qquad (3.7.4)$$

With a single detector, any spatial distribution of u grams is distinguishable from any spatial distribution of v grams if and only if $v > 200u$.

This enormous uncertainty is intolerable, and calls for analysis of a two detector assay system. Since the anterior and posterior detectors are symmetrically equivalent with respect to the simple model we have adopted for the respiratory tract, the minimizing coefficients w_1 and w_2 in eq.(3.2.13) are equal. Thus the relative mass resolution is given by

$$z = \underset{x,y \in X}{\text{maximum}} \frac{f_1(x) + f_2(x)}{f_1(y) + f_2(y)} \qquad (3.7.5)$$

Employing eqs.(1) and (2), the relative mass resolution is readily shown to be

$$z = \cosh\left(\frac{b}{2}\right) = 7.1 \qquad (3.7.6)$$

Thus a pair of detectors placed on opposite surfaces of the patient provides much better resolution than the single-detector design. However a relative mass resolution of 7.1 is still quite large, a fact which results from the assumption that the set of allowed spatial distributions of the aerosol is unconstrained.

3.7.3 Constrained Spatial Distributions

In order to achieve a physically meaningful design it is necessary to account for physical and physiological constraints on the set of spatial distributions of the aerosol. The distribution of the aerosol in the respiratory tract may assume many different forms, depending on the numerous factors which we identified in section 3.7.1. For example, if the particle size is less than several microns a significant portion of the aerosol will penetrate into the pulmonary region. Furthermore, if lung clearance mechanisms are active the aerosol distribution in the pulmonary region will be quite nonuniform. On the other hand, if the particle size exceeds about five microns then most of the aerosol deposition will be restricted to the nasopharyngeal region. In a simple model representing the spatial distribution of small particles (less than one micron) a short time after inhalation (before significant lung clearance occurs), we may assume that the particles are randomly and independently distributed as a cloud throughout the pulmonary region. This is the model we shall study here.

As before, the spatial domain X of the pulmonary region is represented by the unit interval $[0,1]$, corresponding to the posterior-anterior axis. Let us divide X into N equal sub-intervals. The aerosol particles are assumed to be identical, and their total number is M. The probability of any particular particle falling in any given sub-interval is $1/N$, and the total number of particles in the i-th sub-interval is m_i. A given spatial distribution is represented by the N-tuple $(m_1, \ldots, m_N)$. The probability of a particular spatial distribution is given by the multinomial distribution [16]:

$$p(m_1, \ldots, m_N) = \frac{M!}{m_1! \cdots m_N!} \left(\frac{1}{N}\right)^M \tag{3.7.7}$$

The number of particles in the i-th sub-interval is a random variable. The average value of m_i and the covariance of m_i and m_j are

$$E(m_i) = \frac{M}{N} \tag{3.7.8}$$

$$\mathrm{cov}(m_i, m_j) = E(m_i m_j) - E(m_i)E(m_j)$$
$$= \begin{cases} -\frac{M}{N^2} & \text{if } i \neq j \\ \frac{M(N-1)}{N^2} & \text{if } i = j \end{cases} \tag{3.7.9}$$

We wish to distinguish these ensemble-average values of the variables m_i from the spatial average of a given distribution $(m_1, \ldots, m_N)$. This spatial average of a given distribution is the average position or centroid of the aerosol cloud:

$$\mu_1 = \frac{1}{M} \sum_{i=1}^{N} i m_i \tag{3.7.10}$$

This centroid of a given aerosol cloud also has an ensemble average:

$$E(\mu_1) = \frac{1}{M} \sum i E(m_i) = \frac{N+1}{2} \tag{3.7.11}$$

If we subdivide the unit interval into a very large number of sub-intervals, the ensemble average of the aerosol centroid becomes

$$\lim_{N \to \infty} E(\mu_1) = \frac{N}{2}$$

We may express this ensemble average in terms of distance along the unit interval rather than in terms of sub-interval number. We find that the ensemble average of the aerosol centroid position is

$$\lim_{N \to \infty} E(\mu_1) = \frac{1}{2} \qquad (3.7.12)$$

Since the particles are randomly and independently distributed in the interval $[0, 1]$, the ensemble-average of their mean position is the midpoint of the interval.

In order to develop a model for the range of allowed spatial distributions of the aerosol, we wish to evaluate the ensemble variance of the centroid of the aerosol cloud. That is, we wish to evaluate

$$\text{var}(\mu_1) = E\left((\mu_1 - E(\mu_1))^2\right) \qquad (3.7.13)$$

Employing eqs.(8) and (9) this is readily found to be

$$\text{var}(\mu_1) = \frac{N^2 - 1}{12M} \qquad . \qquad (3.7.14)$$

By considering a very large number of sub-intervals, and expressing the variance in terms of distance along the unit interval rather than in terms of the number of sub-intervals, we obtain

$$\lim_{N \to \infty} \text{var}(\mu_1) = \frac{1}{12M} \qquad . \qquad (3.7.15)$$

This expression allows us to estimate how much the centroid of the aerosol cloud will vary from patient to patient. The maximum permissible lung burden (16 nCi) of plutonium oxide aerosol particles of about one micron diameter contains approximately 5000 particles, depending on the actual distribution of sizes. The following table shows the ensemble standard deviation of the centroid of the aerosol for several values of the number of particles.

M	1000	2000	3000	4000	5000
$\sigma(\mu_1)$	0.0091	0.0065	0.0053	0.0046	0.0041

Thus the centroid of a cloud of 1000 particles independently and
randomly distributed on the interval $[0, 1]$ shows an ensemble variation of 0.0091, or almost 1%. Likewise, the centroid of 5000 particles varies by almost 1/2%. This information forms the basis of
our construction of the set of allowed spatial distributions of the
aerosol.

The distributions of a large number of particles on the unit
interval will invariably be small deviations from a uniform distribution. Let $r(x)$ be the number density of particles at point x.
That is, $r(x)\,dx$ is the number of aerosol particles in the interval
$(x, x + dx)$. The uniform number density is simply

$$r(x) = M \qquad\qquad (3.7.16)$$

The set of all allowed distributions comprises small deviations from
eq.(16). Let the set of all allowed spatial distributions of one unit
(in mass) of plutonium aerosol be

$$\widetilde{R}(0,1) = \left\{ r : r(x) = M + \sum_{j=1}^{J} a_j x^j \quad , \quad (a_1, \dots, a_J) \in A \right\}$$

$$(3.7.17)$$

where M is the average number of aerosol particles composing one
unit of mass, and A is a set of J-tuples of small real numbers.

The number of particles in a given spatial distribution $r(x)$ is

$$\mu_0 = \int_0^1 r(x)\,dx = M + \sum \frac{a_j}{j+1} \qquad\qquad (3.7.18)$$

Thus the number of particles varies slightly from distribution to
distribution (in accordance with the actual size-distribution of the
particles). However, since the ensemble-averaged number of particles is M, let the coefficients a_j vary symmetrically and independently around the origin.

To determine the extent of variation of the coefficients a_j, let
us consider the centroid of the spatial distribution $r(x)$:

$$\mu_1 = \frac{1}{M} \int_0^1 xr(x)\, dx = \frac{1}{2} + \frac{1}{M} \sum \frac{a_j}{j+2} \qquad (3.7.19)$$

The ensemble average of the centroid of the aerosol cloud, $E(\mu_1)$, is $1/2$ as it should be. The ensemble variance of the centroid is

$$\mathrm{var}(\mu_1) = E\left(\left(\mu_1 - E(\mu_1) \right)^2 \right)$$

$$= \frac{1}{M^2} \sum_i \sum_j \frac{\mathrm{cov}(a_i, a_j)}{(i+2)(j+2)} \qquad (3.7.20)$$

$$= \frac{1}{M^2} \sum_j \frac{E(a_j^2)}{(j+2)^2} \qquad (3.7.21)$$

since the coefficients vary independently with means equal to zero. Now the extent of variation of the coefficients is established by equating eqs.(15) and (21). Specifically, let A be the set of real J-tuples defined by

$$A = \left\{ a : \sum_{j=1}^{J} \left(\frac{a_j}{j+2} \right)^2 \le \frac{M}{12} \right\} \quad . \qquad (3.7.22)$$

Now it is evident that the set $\widetilde{R}(0,1)$ defined by eqs.(17) and (22) can be represented as the convex hull of the following set of fundamental distributions:

$$D = \left\{ r : r(x) = M + \sum_{j=1}^{J} a_j x^j \quad , \quad (a_1, \ldots, a_J) \in A' \right\} \qquad (3.7.23)$$

where

$$A' = \left\{ a : \sum_{j=1}^{J} \left(\frac{a_j}{j+2} \right)^2 = \frac{M}{12} \right\} \qquad (3.7.24)$$

Let us consider a one-detector system, for which the response function is given by eq.(1). Then the response to a spatial distribution r whose coefficients are $a = (a_1, \ldots, a_J)$ is

$$c(a) = \int_0^1 r(x) f_1(x)\, dx \qquad . \qquad (3.7.25)$$

After some manipulation we find that

$$c(a) = \frac{KM}{b}\left(1 - e^{-b}\right) + K \sum_{j=1}^{J} \frac{j!\, a_j}{b^{j+1}} S_j \qquad (3.7.26)$$

where

$$S_j = 1 - e^{-b} \sum_{i=0}^{j} \frac{b^i}{i!} \qquad (3.7.27)$$

The relative mass resolution is given by the ratio of the maximum to the minimum values of $c(a)$ on the set A'. Explicitly,

$$z = \operatorname*{maximum}_{a,a' \in A'} \frac{c(a)}{c(a')} \qquad (3.7.28)$$

To find the extrema of $c(a)$ we must seek the extrema of

$$T(a) = \sum_{j=1}^{J} \frac{j!\, a_j}{b^{j+1}} S_j \qquad (3.7.29)$$

subject to the constraint:

$$\sum_{j=1}^{J} \left(\frac{a_j}{j+2}\right)^2 = \frac{M}{12} \qquad (3.7.30)$$

We can solve this optimization problem by adjoining the constraint to T with a LaGrange multiplier y as

$$T' = T + y \left(\sum \left(\frac{a_j}{j+2} \right)^2 - \frac{M}{12} \right) \qquad (3.7.31)$$

We must solve the following J equations

$$\frac{\partial T'}{\partial a_j} = 0 \quad , \quad j = 1, 2, \ldots, J \qquad (3.7.32)$$

together with eq.(30) for the $J + 1$ unknowns $a_1, \ldots, a_J$ and y. The maximizing values of the coefficients are found to be

$$\widehat{a}_j = \frac{(j+2)^2 j!\, S_j}{2 b^{j+1}} \left(\frac{3}{M} \sum_{i=1}^{J} \frac{(i+2)i!\, S_i}{b^{i+1}} \right)^{-1/2} \qquad (3.7.33)$$

for $i = 1, 2, \ldots, J$. The minimizing coefficients are $\check{a}_j$ where

$$\check{a}_j = -\widehat{a}_j \quad , \quad j = 1, 2, \ldots, J \qquad (3.7.34)$$

The relative mass resolution now becomes

$$z = \frac{M\left(1 - e^{-b}\right) + \sum\limits_{j=1}^{J} \frac{j!\,\widehat{a}_j}{b^j} S_j}{M\left(1 - e^{-b}\right) - \sum\limits_{j=1}^{J} \frac{j!\,\widehat{a}_j}{b^j} S_j} \, . \qquad (3.7.35)$$

In the following table we show the values of the relative mass resolution z for a range of particle numbers, M. These calculations are quite insensitive to the choice of the highest-order polynomial. The results vary by less than one percent as J varies from 1 to 10. In these calculations $J = 10$.

Relative Mass Resolution for Varying Particle Number

M	1	2	4	7	10
z	1.47	1.31	1.21	1.15	1.13

M	20	40	60	80	100
z	1.089	1.062	1.050	1.043	1.039

M	200	600	1000	2000	5000
z	1.027	1.016	1.012	1.0085	1.0054

From this table we see that for 1000 particles the relative mass resolution is 1.012, indicating a spatial uncertainty of the assay of 1.2%. This is certainly quite adequate in light of the fact that the statistical uncertainty is likely to be greater. Furthermore, only for very small numbers of particles does the spatial uncertainty become appreciable: 5% for 60 particles, 15% for 7 particles and 47% for a single particle. We have noted earlier that several nCi of plutonium aerosol is likely to comprise several thousand particles. Our conclusion must therefore be that, if the particles are of fairly uniform size and if they are distributed randomly and independently in the pulmonary region with equal probability of deposition in any area, then a single anterior detector is adequate for overcoming the spatial uncertainty of the measurement. We must stress that a number of factors could lead to violation of these conditions. If the particle size exceeds several microns then the major portion of the aerosol will be deposited in the nasopharyngeal and tracheobronchial regions. Also if lung clearance mechanisms have operated (for which probably only a few hours are adequate) then the aerosol distribution may be far from uniform in the pulmonary or other regions of the respiratory tract. If either of these mechanisms may be expected, then the design must be based on analysis of a much wider range of allowed spatial distributions of the aerosol. One should then expect larger values of the relative mass resolution. We shall return to this topic in section 6.4.

3.8 EXAMPLE: THICKNESS MEASUREMENT

3.8.1 Formulation

The measurement of thickness by absorption of gamma- or X-radiation is a very widespread technique. The fractional transmission of radiation upon passing through the sample is used to deduce the average thickness in the irradiated region. This approach has been employed in measuring the thickness of a layer of coal or other material on a conveyor belt [17], or the density and composition of multicomponent fluids or slurries in hydraulic transport in pipes [18], or in measuring the thickness of thin foils [19].

In all of these applications the sample is likely to display variable linear or density thickness across the irradiated region. In the case of thin foils this may arise from the rolling process by which the foil is manufactured or from non-uniform chemical decomposition of the foil. Conveyor belts may display non-uniform thickness profiles due to irregular composition, loading or settling of the transported material. Multicomponent slurries or fluids may show non-uniform profile of the density thickness due to the random spatial distribution of the phases in the transport channel. Calibration, on the basis of a uniform thickness profile, of the intensity of the transmitted radiation for samples with variable thickness has been shown to lead to unacceptable results [20]. That is, different thickness profiles lead to appreciably different fractional transmissions, even when the average thickness is the same.

The instrumentalist must design his assay system so as to maximize the capability to resolve different regionally-averaged thicknesses, in the light of a certain range of variability of the thickness profiles. Typically the designer is free to optimize his choice of the energy of the transmitted radiation and of the duration of the measurement. He must choose his design so as to minimize the variation of response to different profiles with the same average thickness. For a given application one is likely to be able to esti-

mate the typical average thickness and the range of variation of
the thickness across the sample. This information may be used
to define the set of allowed spatial distributions (thickness profiles). The design can then be optimized by employing the results
of section 3.3 for constrained spatial distributions.

In this section we shall consider the optimization of the energy
of the transmitted radiation. After discussing the general formulation we shall present numerical results for coal on a conveyor
belt.

We shall represent the thickness profiles by expanding them
in a series of orthogonal functions. The expansion coefficients will
be allowed to vary in order to represent the range of variability
of the thickness profile. The thickness at point x in the interval
$[0, L]$ is represented by

$$r(x) = \sum_{j=0}^{J} a_j s_j \left(\frac{x}{L} \right) \qquad (3.8.1)$$

We may choose the s_j to be trigonometric functions or orthogonal
polynomials. However, by choosing the s_j to be Rademacher functions [21] we will be able to reach a completely analytical result.

The Rademacher functions may be defined on the unit interval $[0, 1]$ as follows. For any real number x, let $[x]$ represent the
greatest integer less than or equal to x. Then the j-th Rademacher
function is

$$s_j(x) = (-1)^{[x2^j]} \qquad , \qquad j = 0, 1, 2, \ldots \qquad (3.8.2)$$

These functions are square-waves as shown in figure 3.8.1. The
Rademacher functions are orthonormal. That is

$$\int_0^1 s_j(x) s_k(x)\, dx = \begin{cases} 0 & \text{if } j \neq k \\ 1 & \text{if } j = k \end{cases} \qquad (3.8.3)$$

Throughout this section we shall consider samples of integral
length L. The set of all allowed thickness profiles on the interval
$[0, L]$ is

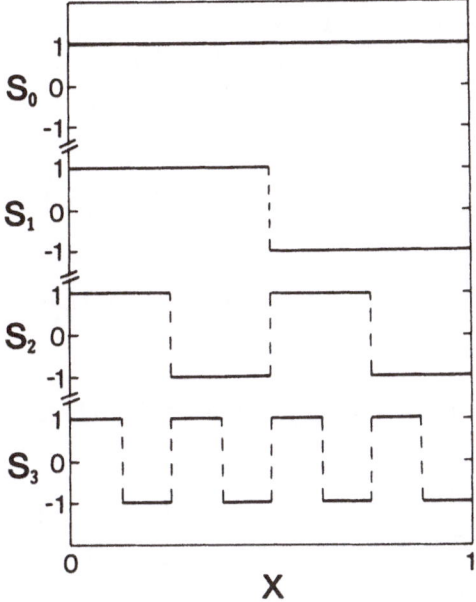

Figure 3.8.1   The first four Rademacher functions.

$$\widetilde{R} = \left\{ r : r(x) = \sum_{j=0}^{J} a_j s_j \left(\frac{x}{L}\right) \quad, \quad (a_0, \ldots, a_J) \in A \right\} \quad (3.8.4)$$

where A is a set of $(J+1)$-tuples which defines the range of variability of the thickness profiles. The set A may be defined in various ways depending on what information is available on the application in question. Usually one is able to estimate the mean and variance of the thickness profile, and the set A should be chosen so that $\widetilde{R}$ matches this information. For a given profile $r(x)$, the mean thickness on an interval of integral length L is

$$E(r) = \frac{1}{L} \int_0^L r(x) \, dx$$

$$= \frac{1}{L} \sum_{0}^{J} a_j \int_{0}^{L} s_j \left(\frac{x}{L}\right) dx$$

$$= a_0 \tag{3.8.5}$$

Likewise the variance of the thickness of the profile $r(x)$ is

$$\mathrm{var}(r) = \mathrm{E}(r^2) - \mathrm{E}(r)^2 = \sum_{1}^{J} a_j^2 \tag{3.8.6}$$

In light of eqs.(5) and (6), a reasonable choice [22] of the set A is

$$A = \left\{ a : (a_0 - T)^2 + \sum_{1}^{J} a_j^2 \leq Q \right\} \tag{3.8.7}$$

where T and Q are known quantities.

The set of allowed profiles of average thickness u on an interval of integral length L is given by

$$\tilde{R}(1, u) = \left\{ r : r(x) = \sum_{0}^{J} a_j s_j \left(\frac{x}{L}\right) \quad , \quad (a_0, \ldots, a_J) \in A(1, u) \right\} \tag{3.8.8}$$

where

$$A(1, u) = \left\{ a : \overset{\cdot}{a_0} = u \quad , \quad \sum_{1}^{J} a_j^2 \leq Q - (T - u)^2 \right\} \tag{3.8.9}$$

Some care must be taken in the choice of T, Q and J, as we shall show, to assure that all the profiles are non-negative throughout the interval $[0, L]$.

The intensity of radiation transmitted through a given profile $r(x)$ defined by the coefficients $a_0, \ldots, a_J$ is

$$c(a) = Kt \int_0^L e^{-\mu r(x)} \, dx \qquad (3.8.10)$$

where K is a constant and t is the duration of irradiation of each unit length of the profile. Employing the explicit expressions for the Rademacher functions in the expansion of $r(x)$, eq.(10) becomes

$$c(a) = KLt \int_0^1 \exp\left(-\mu \sum_{j=0}^J a_j(-1)^{[x2^j]}\right) dx \qquad . \qquad (3.8.11)$$

The highest order Rademacher function, $s_J(x/L)$, has 2^J lobes in $[0, 1]$. These lobes coincide with the following intervals:

$$I(k) = \left[\frac{k}{2^J}, \frac{k+1}{2^J}\right] \quad , \quad k = 0, 1, 2, \ldots, 2^J - 1 \qquad (3.8.12)$$

So eq.(11) becomes

$$c(a) = KLt \sum_{k=0}^{2^J-1} \int_{I(k)} \exp\left(-\mu \sum_{j=0}^J a_j(-1)^{[x2^j]}\right) dx \qquad (3.8.13)$$

The distinctive feature of the Rademacher functions is that they are all constant (for $j \leq J$) in each of the intervals $I(k)$. This confers a particular advantage for the present application to the Rademacher functions over trigonometric functions or orthogonal polynomials [23]. For trigonometric functions or polynomials the integrals in eq.(13) are intractable, while by exploiting the square-wave form of the Rademacher functions one obtains

$$c(a) = \frac{KLte^{-\mu a_0}}{2^J} \sum_{k=0}^{2^J-1} \exp\left(-\mu \sum_{j=1}^J a_j(-1)^{[k2^{j-J}]}\right) dx \quad (3.8.14)$$

From this relation we see that $c(a)$ consists of the sum of 2^J exponential terms, and that the exponent of each term is the sum of the J coefficients $a_1, \ldots, a_J$ each of positive or negative sign. There are in fact precisely 2^J distinct configurations of $\sum \pm a_j$, and one may readily see that each of these configurations appears just once in eq.(14). Thus we find that

$$c(a) = \frac{KLte^{-\mu a_0}}{2^J} \sum_{m_1=\pm 1} \cdots \sum_{m_J=\pm 1} \exp\left(-\mu \sum_{j=1}^{J} m_j a_j\right)$$

$$(3.8.15)$$

By factoring out common terms in this multiple sum one arrives at the following concise expression for $c(a)$ in terms of hyperbolic cosine functions.

$$c(a) = KLte^{-\mu a_0} \prod_{j=1}^{J} \cosh(\mu a_j) \quad . \qquad (3.8.16)$$

3.8.2 A Minimization Problem: Choosing Q

A given thickness profile $r(x)$ on the interval $[0, L]$ is specified by the coefficients $a_0, \ldots, a_J$ in eq.(1). The smallest value which $r(x)$ assumes in the interval $[0, L]$ occurs when all the Rademacher functions except s_0 are negative. This minimum value is

$$r(x) = a_0 - \sum_{1}^{J} a_j \qquad (3.8.17)$$

We must choose the set A of $(J + 1)$-tuple coefficients so that the profiles $r(x)$ remain non-negative for all allowed values of $a_1, \ldots, a_J$. In order to establish the constraints which this imposes on the choice of T, Q and J, let us find the smallest value $G_J(Q)$ which the right-hand side of eq.(17) may assume, subject to the constraints of the set A. Specifically, we must find

$$G_J(Q) = \underset{a_0,\ldots,a_J}{\text{minimum}} \left(a_0 - \sum_{1}^{J} a_j\right) \qquad (3.8.18)$$

subject to the constraint

$$(a_0 - T)^2 + \sum_1^J a_j^2 = Q \qquad (3.8.19)$$

(Since we are seeking the minimum of a linear functional on the compact convex set A, the minimum will occur on the boundary of A. Hence we are justified in using an equality constraint rather than the inequality appearing in the definition of A).

Rather than directly solving eq.(18), it is convenient to first find the following maximum

$$F_J(x) = \underset{a_1,\ldots,a_J}{\text{maximum}} \sum_1^J a_j \qquad (3.8.20)$$

subject to the constraint

$$\sum_1^J a_j^2 = x \qquad .$$

We shall illustrate the maximization of eq.(20) using the recursive technique known as dynamic programming [24]. Suppose $J = 1$. Then

$$F_1(x) = \underset{a_1}{\text{maximum}}\, a_1 = \sqrt{x} \quad .$$

Now suppose that $J = 2$. For any given choice of the coefficient a_1, the maximum value which $a_1 + a_2$ may assume is $a_1 + F_1(x - a_1^2)$. Hence

$$F_2(x) = \underset{a_1}{\text{maximum}}\left(a_1 + F_1\left(x - a_1^2\right)\right)$$

$$= \underset{a_1}{\text{maximum}}\left(a_1 + \sqrt{\left(x - a_1^2\right)}\right)$$

$$= \sqrt{2x} \quad .$$

We may adopt the inductive hypothesis that for arbitrary J

$$F_J(x) = \sqrt{Jx} \qquad . \qquad (3.8.21)$$

The maximum for $J + 1$ is

$$F_{J+1}(x) = \operatorname*{maximum}_{a_1}\left(a_1 + F_J(x - a_1^2)\right) \quad .$$

The hypothesis is verified by establishing that

$$F_{J+1}(x) = \operatorname*{maximum}_{a_1}\left(a_1 + \sqrt{J(x - a_1^2)}\right)$$
$$= \sqrt{(J + 1)x}$$

Now return to eqs.(18) and (19).

$$G_J(Q) = \operatorname*{minimum}_{a_0}\left(a_0 - F_J(Q - (a_0 - T)^2)\right)$$

where

$$(a_0 - T)^2 \le Q \quad .$$

Employing eq.(21)

$$G_J(Q) = \operatorname*{minimum}_{a_0}\left(a_0 - \sqrt{JQ - J(a_0 - T)^2}\right)$$
$$= T - \sqrt{(1 + J)Q}$$

In order to avoid negative thicknesses $(r(x) < 0)$ we must choose Q, T and J so that

$$G_J(Q) \ge 0$$

or equivalently

$$Q \le \frac{T^2}{J + 1} \quad .$$

3.8.3 Resolution Without Statistical Uncertainty

The complete response set for allowed thickness profiles of average thickness u is

$$\widetilde{C}(1, u) = \{c : c = c(a) \quad , \quad (a_0, \ldots, a_J) \in A(1, u)\}$$

where $c(a)$ is given by eq.(16) and $A(1, u)$ by eq.(9). Since $A(1, u)$ is convex and compact, so is $\widetilde{C}(1, u)$. However, since $c(a)$ is nonlinear in u, we see that the class of sets $\left\{\widetilde{C}(1, u), u > 0\right\}$ does not have the property of proportionality. Consequently we must use Theorem 4 of section 3.3 to establish the range of distinguishable values of the average thickness. Any allowed spatial distribution whose average thickness is u is distinguishable from any allowed spatial distribution whose average thickness is v if and only if either

$$\underset{a \in A(1,u)}{\text{maximum}} c(a) < \underset{a' \in A(1,v)}{\text{minimum}} c(a')$$

or

$$\underset{a \in A(1,v)}{\text{maximum}} c(a) < \underset{a' \in A(1,u)}{\text{minimum}} c(a')$$

To establish the range of distinguishability we must find the extrema of

$$c(a) = KLt\, e^{-\mu a_0} \prod_{j=1}^{J} \cosh(\mu a_j)$$

subject to the constraint

$$a_0 = u \quad , \quad \sum_{j=1}^{J} a_j^2 \le Q - (T - u)^2$$

Since $\cosh(x)$ achieves a minimum $(= 1)$ at $x = 0$, it is evident that the minimum of $c(a)$ on $A(1, u)$ occurs at

$$a_1 = \cdots = a_J = 0 \qquad .$$

Thus the minimum of $c(a)$ on $A(1, u)$ is

$$\check{c}(u) = KLt\, e^{-\mu u}$$

We shall proceed with a dynamic programming approach to find the maximum value of $c(a)$. Let

$$V = Q - (T - u)^2$$

represent the quantity defining the constraint. If $J = 1$, then the maximum is

$$F_1(V) = \operatorname*{maximum}_{a_1} \cosh(\mu a_1) = \cosh\left(\mu\sqrt{V}\right) \qquad .$$

If $J = 2$, then the maximum is

$$F_2(V) = \operatorname*{maximum}_{a_1}\left(\cosh(\mu a_1)F_1\left(V - a_1^2\right)\right)$$

$$= \left(\cosh\left(\mu\sqrt{\frac{V}{2}}\right)\right)^2$$

We may adopt the inductive hypothesis that

$$F_J(V) = \left(\cosh\left(\mu\sqrt{\frac{V}{J}}\right)\right)^J \qquad .$$

This hypothesis is verified by showing that, together with the previous equation,

$$F_{J+1}(V) = \operatorname*{maximum}_{a_1}\left(\cosh(\mu a_1)F_J\left(V - a_1^2\right)\right)$$

$$= \left(\cosh\left(\mu\sqrt{\frac{V}{J+1}}\right)\right)^{J+1} \qquad .$$

Consequently the maximum response obtainable from a profile of average thickness u, is

$$\widehat{c}(u) = KLt\,e^{-\mu u}\left(\cosh\left(\mu\sqrt{\frac{Q-(T-u)^2}{J}}\right)\right)^J$$

The values u and v of the average profile-thickness are distinguishable if and only if the intevals $[\check{c}(u),\widehat{c}(u)]$ and $[\check{c}(v),\widehat{c}(v)]$ are disjoint.

Let us now determine the range of distinguishable average thickness for the assay of coal on a conveyor belt. The coal layer in our application has a typical thickness of 40 cm. Our aim is to compare the resolution capability of radiations of different energies. The radiation sources which are available are

Source	Gamma Energy (MeV)	Approximate μ (1/cm)
^{60}Co	1.17 , 1.33	0.06
^{137}Cs	0.662	0.09
^{241}Am	0.0595	0.2

The available gamma energies range from 59.5 keV to 1.33 MeV. The values of the linear absorption coefficients depend on the precise composition and compactness of the coal. Approximate values of μ are shown in the table. We shall examine two cases, $\mu = 0.06$ and $\mu = 0.2$ 1/cm. Our analysis is based on expanding the thickness profiles in 21 Rademacher functions. That is, $J = 20$. We shall consider two different degrees of variability of the profiles: $Q = 9.52$ cm^2 and $Q = 38.1$ cm^2.

In figure 3.8.2 we show the minimum and maximum values of the response versus the average thickness of the profile. In this figure, $Q = 9.52$ cm^2. The responses have been normalized to the maximum value of $\widehat{c}$. For the smaller absorption coefficient

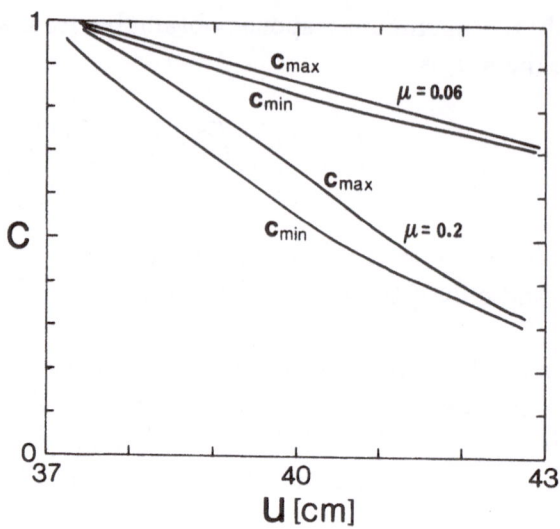

Figure 3.8.2 Minimum and maximum normalized response versus the average thickness of the profile. T=40 cm, Q=9.52 cm^2, J=20.

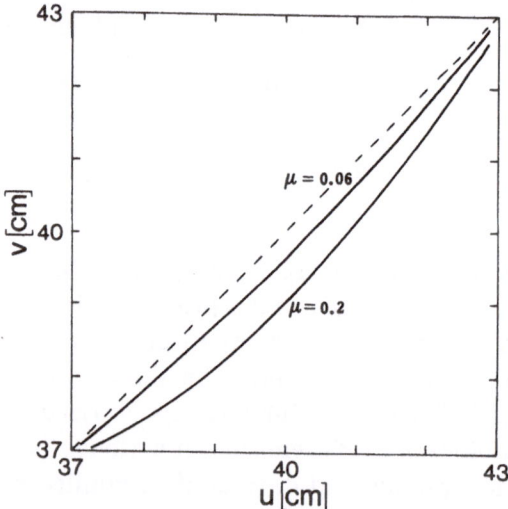

Figure 3.8.3 Lines of distinguishability. T=40 cm, Q=9.52 cm^2, J=20.

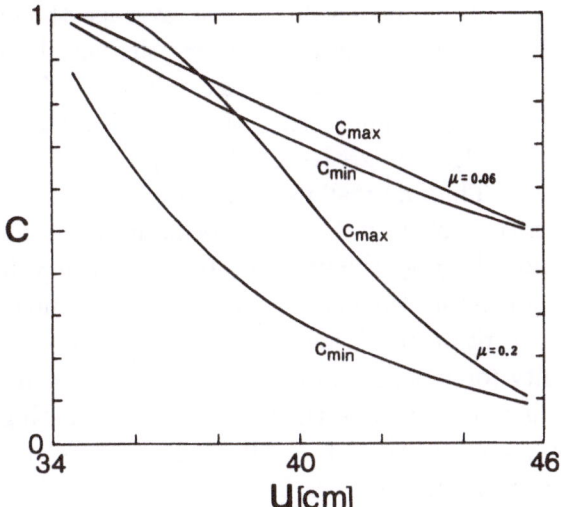

Figure 3.8.4 Minimum and maximum normalized response versus the average thickness of the profile. $T=40$ cm, $Q=38.1$ cm^2, $J=20$.

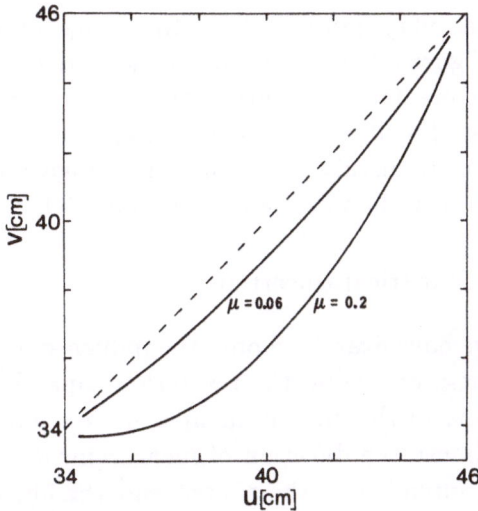

Figure 3.8.5 Lines of distinguishability. $T=40$ cm, $Q=38.1$ cm^2, $J=20$.

(higher gamma energy) the range of variability of the response is comparatively small for fixed value of the average thickness. At the larger value of μ (lower gamma energy) the variability of the response is greater. Thus, for an average thickness of 40 cm, the minimum and maximum normalized responses are 0.55 and 0.67 respectively ($\mu = 0.2$). The greatest value of u for which $\check{c}$ exceeds 0.67 is 39.1 cm. Thus any allowed distribution whose average thickness is less than 39.1 cm is distinguishable from any allowed distribution whose average thickness is 40 cm. In this way the full range of distinguishable thicknesses may be established for both values of the absorption coefficient. Figure 3.8.3 shows the lines of distinguishability for both gamma energies. Each line defines the greatest value v of the average thickness which is distinguishable from the corresponding value of u. All smaller values of v are also distinguishable from the same value of u. The fact that the line for $\mu = 0.06$ is above the line for $\mu = 0.2$ indicates the superior resolution obtained at the higher gamma energy.

Figure 3.8.4 shows the normalized values of the minimum and maximum responses for more variable profiles: $Q = 38.1$ cm^2. Since the profiles vary more widely, the range of responses is greater than in figure 3.8.2. Figure 3.8.5 shows the lines of distinguishability. The greater variability of the profiles is manifested in poorer resolution. For instance, for $\mu = 0.2$, the greatest average thickness which is distinguishable from an average thickness of 40 cm is 36.2 cm, rather than 39.1 cm as in figure 3.8.3.

3.8.4 Including Statistical Uncertainty

Up to now we have examined only the influence of the variability of the thickness profile on the resolution capability. The statistical uncertainty of the measurement is no less important. The designer must choose the duration of measurement, the length of the sample, the intensity of the source and the efficiency of the detector. Proper choice of these design parameters requires consideration of both the spatial and the statistical uncertainties.

The results of section 3.3.3 may be applied directly in this example. If $v < u$, then every allowed spatial distribution whose average thickness is v is distinguishable, to a specified level of statistical confidence, from every allowed spatial distribution whose average thickness u if and only if

$$\check{c}(v) - b\sigma\big(\check{c}(v)\big) > \hat{c}(u) + b\sigma\big(\hat{c}(u)\big)$$

where $\sigma(c)$ is the statistical standard deviation of the measurement c, and b is a non-negative number which determines the level of statistical confidence.

3.9 EXAMPLE: ENRICHMENT ASSAY

We shall now consider an example in which the design-analysis is applied to a rather different sort of measurement. Let us suppose that we wish to determine the enrichment or composition of one species which is homogeneously blended into another species. This mixture, which is a powdered solid, is in a closed container and is unavailable for direct observation. We are unable to determine the distribution of the solid on the bottom of the container. The solid may cover the entire bottom surface of the container, or it may cover only a fraction of this surface, leaving a fraction s empty. Furthermore the boundary of the distribution of the solid is completely variable.

This type of assay problem arises in the determination of the enrichment of ^{235}U in uranium hexafluoride. For the sake of clarity and concreteness we shall continue in the context of this particular application. Our measurement is performed by placing a detector on the top of the container. The detector senses the 186 keV gamma radiation from the ^{235}U. We shall assume that the solid, wherever it is present, forms a layer which is effectively infinitely thick with respect to the 186 keV gamma radiation. Over a reasonable range of enrichments the response per unit area is proportional to the enrichment. However, since the actual area covered by the solid is unknown, we have a definite uncertainty in the calibration.

The detector views the entire bottom of the container, and the detector response varies as a point-source is moved on this surface.

For simplicity we shall assume that the detector response varies with only one degree of freedom: the deviation from the center of the field of view of the detector. The point-source response function for unit enrichment is

$$f(x) = b e^{-\mu x} \quad , \qquad -D \le x \le D \qquad (3.9.1)$$

The fraction s of the empty surface may take any in the range of values from 0 to S. Let us suppose that we are able to estimate S. The greatest response of the detector occurs when the surface is entirely covered by analyte material, and is given by

$$\hat{c} = 2 \int_0^D f(x)\, dx = \frac{2b}{\mu} \left(1 - e^{-\mu D} \right) \quad . \qquad (3.9.2)$$

The least possible response occurs when the fraction S of the surface is empty, and when this void is symmetrically located in the center of the field of view of the detector. Thus the minimum response is

$$\check{c} = 2 \int_{\frac{SD}{2}}^D f(x)\, dx = \frac{2b}{\mu} \left(\exp\left(-\frac{S\mu D}{2} \right) - \exp(-\mu D) \right) \qquad (3.9.3)$$

The relative enrichment resolution with one detector is the ratio $\hat{c}/\check{c}$. If we let $y = \exp(-\mu D)$ we find the relative resolution to be

$$z(1) = \frac{1 - y}{y^{S/2} - y} \quad . \qquad (3.9.4)$$

From this relation we see that uncertainty in the resolution is introduced by two factors. The first factor is the unknown fractional surface which is unoccupied by the analyte (with a maximum value of S). The second factor is the degree of field-position sensitivity of the detector as expressed by $y = \exp(-\mu D)$. If the

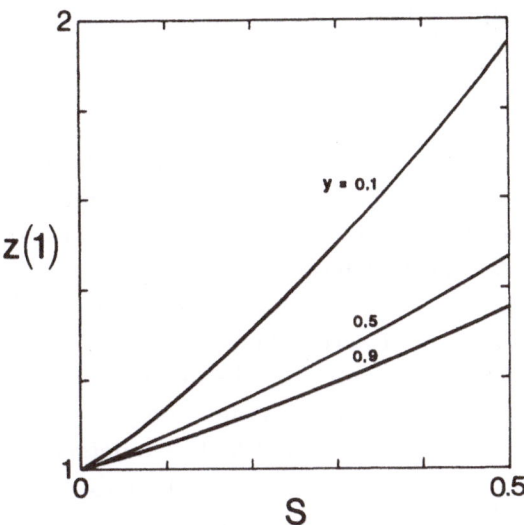

Figure 3.9.1 Relative enrichment resolution for one detector, $z(1)$, versus the maximum fractional void S, for various degrees of spatial sensitivity of the detector y.

detector is entirely insensitive to the source position in its field of view ($\mu = 0$), we still have uncertainty resulting from the first factor. This limiting uncertainty is

$$\lim_{\mu \to 0} z(1) = \frac{1}{1 - S/2} \quad .$$

In figure 3.9.1 we show the relative enrichment resolution versus the value of the maximum void S, for various degrees of position-sensitivity of the detector. When the detector response varies strongly with position ($y = 0.1$) the relative resolution reaches a value of nearly 2 at a maximum void of 50% ($S = 0.5$). At the other extreme, when the detector response is nearly flat ($y = 0.9$) the relative resolution at a maximum void of 50% is 1.35.

If the results of the analysis up to this point indicate an unacceptably large value of the relative enrichment resolution for the values of S and y which we anticipate, we should examine designs employing multiple measurements.

Let us first consider two measurements, each of which views just half of the bottom surface of the container. For each measurement the response function is the same as before (eq.(1)) but now x for each detector varies over the smaller interval $[-D/2, D/2]$. Thus the advantage gained is that the response function of each measurement varies less over its field of view. However each individual measurement is now more sensitive to the unknown fraction of empty surface. This comes about because one of the measurements may experience a fractional empty surface as large as $2S$ (if $S \leq 1/2$) if all the empty surface falls in the field of view of that measurement. By employing the design analysis we shall obtain a quantitative assessment of just how well two detectors perform as compared with one detector. In fact, we shall obtain an analytical expression for the relative enrichment resolution of an arbitrary number of detectors.

Our first task is to construct the set of all pairs of responses — the complete response set — for any allowed fractional void and any spatial distribution of the analyte material. The complete response set for unit enrichment, $C(1)$, is shown in figure 3.9.2 for $S = y = 0.5$. The point A occurs when both detectors view zero void (completely covered surfaces). The points along the segment AB occur when detector 1 views a completely covered surface and detector 2 views various degrees of void. Similarly for the segment AB'. The arc from B to B' arises when the maximum void, S, is distributed variously between the two detectors, and when each detector views a spatial distribution with the least possible response for that value of void. We must bear in mind that when say 30% of the maximum void S is in the field of view of detector 1, the fractional void for that detector is in fact $2 \times 0.3 \times S$. This is because the field of view of one detector is only half of the combined fields of view of the two detectors. Thus when a fraction g of the maximum void S occurs in the field of view of detector 1, the minimum response of the detector is found from eq.(3) by replacing D by $D/2$ and S by $2gS$:

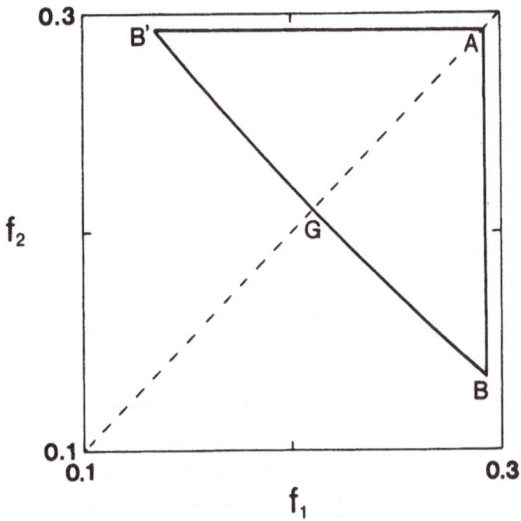

Figure 3.9.2 Complete response set for two detectors. $S=y=0.5$. The responses have been normalized by the factor $\mu/2b$.

$$f_1 = \frac{2b}{\mu}\left(\exp\left(-\frac{gS\mu D}{2}\right) - \exp\left(-\frac{\mu D}{2}\right)\right) \qquad (3.9.5)$$

At the same time the response of detector 2 is

$$f_2 = \frac{2b}{\mu}\left(\exp\left(-\frac{(1-g)S\mu D}{2}\right) - \exp\left(-\frac{\mu D}{2}\right)\right) \qquad (3.9.6)$$

The arc BB' is defined by the locus of points (f_1, f_2) as g varies from 0 to 1. It can be shown that this arc is convex, so the boundary of the complete response set is convex.

Points interior to the boundary in figure 3.9.2 are generated by other values of the total void and by various spatial distributions of the analyte material.

Since the detector response is proportional to the enrichment E, we see that the class of complete response sets $\{C(E),\ E \geq 0\}$ has the property of proportionality. Furthermore since $C(1)$ is convex and symmetric, the expansion of $C(1)$ is given by the ratio of point A to point G in figure 3.9.2. Thus $z(2)$ is found to be

$$z(2) = \frac{1 - y^{1/2}}{y^{S/4} - y^{1/2}} \tag{3.9.7}$$

where as before $y = \exp(-\mu D)$. Comparison of eqs.(4) and (7) shows that the resolution with two detectors is always at least as good as with one. That is, $z(2) \leq z(1)$. In the following table we present the ratio $z(2)/z(1)$ versus the value of S, the maximum fractional void of the surface, for various values of y. These calculations show that two detectors provide only very slight improvement over a single detector, except when the response is very non-uniform (y small). We may feel that the resolution with one detector is not very satisfactory. However the results of our quantitative design analysis show that two detectors usually will provide almost no improvement.

S	$z(2)/z(1)$		
	$y = 0.9$	$y = 0.5$	$y = 0.1$
0.0	1	1	1
0.1	0.999	0.990	0.958
0.2	0.997	0.980	0.918
0.3	0.996	0.970	0.880
0.4	0.995	0.961	0.844
0.5	0.994	0.951	0.810

There is no difficulty in extending our analysis to the case of n identical detectors. Each detector views a $1/n$-th part of the total field of view and has the response function of eq.(1) over the interval $[-D/n, D/n]$, where x is taken relative to the center of the field of view of the detector. We may construct the complete response set for n detectors as an n-dimensional analog of the 2-dimensional set in figure 3.9.2. We will find the same symmetry properties. Thus the relative enrichment resolution for n detectors will be the ratio of the response of one detector to a fully covered surface, to the minimal response when $1/n$-th of the maximal void S is present in each field of view. When $1/n$-th of the void is present in a field which is $1/n$-th of the total field, the local void fraction is S. Thus the relative enrichment resolution with n detectors is

$$z(n) = \frac{1 - y^{1/n}}{y^{S/2n} - y^{1/n}} \tag{3.9.8}$$

where as before $y = \exp(-\mu D)$, which represents the spatial variation of the response of a single detector over the entire field of view.

Several points should be noted in connection with this expression. First of all, when the response function is perfectly flat $(\mu = 0)$, we find that $z(n)$ becomes independent of n. That is

$$\lim_{\mu \to 0} z(n) = \frac{1}{1 - S/2} \quad . \tag{3.9.9}$$

In other words, when the response is entirely insensitive to the position of the source material, no advantage is gained by performing successive measurements by subdivision of the total field of view. This is true because our design is deterministic — we are not considering the possibility of a probabilistic interpretation of measurement. Indeed we would have to augment our formulation of this assay problem with considerable additional information in order to enable a probabilisitc analysis. In addition, the design would undoubtedly be much more involved. Probabilistic considerations are deferred to Chapters 4 and 5.

Eq.(8) shows an additional interesting property. Regardless of the value of μ, the relative resolution converges for large n to the expression in eq.(9):

$$\lim_{n \to \infty} z(n) = \frac{1}{1 - S/2} \quad . \tag{3.9.10}$$

In other words, the uncertainty introduced by non-flatness of the response function may be overcome by employing a sufficiently large number of detectors. This is not surprising intuitively since, for large n, each detector "sees" essentially only a single point on the surface of the analyte. The importance of eq.(8) is that it provides a quantitative assessment of how quickly this asymptotic resolution is reached.

In the following table we show the relative enrichment resolution for various values of y and n. $S = 0.5$ throughout. The limiting value of the resolution is 1.333. Now we are able to appreciate that, when $y = 0.9$, the relative resolution with a single detector is only 1.3% above the asymptotc value. Consequently, additional measurements are unable to appreciably improve the resolution. When $y = 0.5$, the relative resolution with a single detector is 10% above the asymptotic value; with four detectors the resolution is 2.2% above the asymptotic value. When $y = 0.1$, the resolution with one detector is 46% greater than the asymptotic value; with four detectors the resolution is 8.1% above the asymptotic value; with 16 detectors the resolution is 1.9% greater than the limiting resolution.

The conclusion of our analysis is that the limiting relative enrichment resolution given by eq.(10) can be approached (with fewer or more measurements depending on the value of y), but cannot be improved so long as we restrict ourselves to the measurement regime which we defined at the outset. If better resolution in required, we must employ measurements of a different sort — perhaps by positioning detectors at various angles to the container — or we must exploit probabilistic techniques of measurement interpretation.

n		$z(n)$	
	$y = 0.9$	$y = 0.5$	$y = 0.1$
1	1.351	1.467	1.947
2	1.342	1.395	1.577
4	1.338	1.363	1.441
8	1.336	1.348	1.385
16	1.334	1.341	1.359

3.10 AUXILIARY PARAMETERS

We have now established a general tool for optimizing the design of an assay-system which is to measure samples containing spatially-random source material. Our analysis is based on knowledge of the point-source response functions of the detector, and knowledge of the set of allowed spatial distributions of the source material. We have assumed that the spatial structure of the matrix, in which the source material is embedded, is non-varying throughout the ensemble of samples. Furthermore, we have assumed that the point-source response functions are known with arbitrary accuracy. In many situations of practical interest one or more of these assumptions may be unrealistic. In this and the following two sections we shall demonstrate how the design analysis may be readily applied to these situations.

Our treatment of variability of the matrix structure or of the response function may be formulated in a somewhat more general context. Just as the detector-response is sensitive to the spatial distribution of the source material, so is the response sensitive to many *auxiliary parameters*. These parameters may be matrix structure, wall-thickness of the sample container, operating temperature of the detector, voltage of detector power supplies, and so on. Our aim is to evaluate the relative resolution of an arbitrary spatial moment of the source material, in light of variability of specified auxiliary parameters. The importance of such an analysis is two-fold. First, the design can be optimized to confront specified expected ranges of variation of relevant auxiliary parameters. Second, and conversely, this analysis can be used to establish specifications on the range of permissible variation of auxiliary parameters.

This design analysis displays the two basic properties which have characterized our results up to now. First, we do not need to assume small variations of the auxiliary parameters, because the relative resolution is not based on analysis of first-order variations of the response. Second, an efficient computerizable algorithm is obtained which enables ready evaluation of the influence of variation of a large number of interrelated auxiliary parameters.

We shall begin by examining a number of illustrative examples, before addressing some general aspects of the problem.

Consider a sample in which the source material is homogeneously distributed. The response of a single detector to u grams of source material is $f = au$, where the constant a may be determined by calibration with standardized samples. Now suppose the wall-thickness of the sample container, t, may vary from sample to sample, and thereby influence the response to u grams, according to

$$f = ae^{-wt}u \quad , \quad t_1 \leq t \leq t_2$$

where w is known. Thus the set of responses to u grams is

$$\widetilde{C}(0, u) = \left\{ c : c = ae^{-wt}u \quad , \quad t_1 \leq t \leq t_2 \right\}$$

which is convex. Also the class of sets $\left\{ \widetilde{C}(0, u) \, , \, u > 0 \right\}$ has the property of proportionality. Thus the relative resolution for the zeroth moment is the expansion of $\widetilde{C}(0, 1)$:

$$z(0) = \text{e}\left(\widetilde{C}(0, 1) \right) = e^{w(t_2 - t_1)}$$

It is instructive to generalize this example to the case of n measurements. Let the n-vector response to the uniform spatial distribution of u grams of source material be

$$f = aub = au(b_1, \ldots, b_n)$$

where b is an n-vector auxiliary parameter that varies on a set B. The complete response set is

$$\widetilde{C}(0, u) = \{c : c = aub \quad, \quad b \in B\}$$

If B is convex then so is $\widetilde{C}(0, u)$. Also, the class $\left\{\widetilde{C}(0, u), \ u > 0\right\}$ has the property of proportionality. Let us suppose that B is an n-sphere defined by

$$B = \left\{b = (b_1, \ldots, b_n) : \sum_1^n (b_i - b')^2 \leq 1\right\}$$

where b' is known and exceeds unity. As before, the relative mass resolution is just the expansion of $\widetilde{C}(0, 1)$, which equals the expansion of B:

$$z(0) = \mathrm{e}\left(\widetilde{C}(0, 1)\right) = \mathrm{e}(B)$$

Because B is highly symmetrical, its expansion is readily calculated. Consider a ray from the origin which passes through B. Since any two points on such a ray are proportional to one another, one point can be viewed as the expansion of the other. Consider the points on the intersection of this ray and the boundary of B. The constant of proportionality between these two points is precisely the expansion of B along this particular ray. By seeking the ray with the greatest expansion one finds the expansion of B. It is readily shown that, in our example, the expansion of B occurs along the ray defined by $b_1 = \cdots = b_n$. Thus the expansion is

$$z(0) = \frac{\sqrt{n}\, b' + 1}{\sqrt{n}\, b' - 1}$$

The following table shows how the relative resolution improves as the number of measurements increases. Two different values of b' are shown.

Relative Resolution $z(0)$

	$n=1$	2	3	4	6	8	10
$b' = 2$	3.00	2.09	1.81	1.67	1.51	1.43	1.38
$b' = 10$	1.22	1.15	1.12	1.11	1.085	1.073	1.065

The relative resolution steadily improves, though the utility of the marginal detector decreases progressively.

3.11 EXAMPLE: VARIABLE MATRIX STRUCTURE

Variability of the structure of the matrix of the sample can be treated in the context of random auxiliary parameters. Since matrix variability is particularly important we shall discuss an example in some depth.

The response of an ideal detector to a point-source of gamma radiation at a distance x in a homogeneous matrix is

$$f(x) = \frac{k}{d^2(x)} \, e^{-\mu d(x)}$$

where k is a constant relating to detector efficiency, $d(x)$ is the distance from x to the detector, and μ is the linear absorption coefficient of the matrix. If the composition of the matrix varies along the path from the point x to the detector, then the absorption coefficient is likely to vary as well. In this case, the response to a point-source at position x is

$$f(x) = \frac{k}{d^2(x)} \, \exp\left(-\int \mu(x') \, dx'\right)$$

where the integral is along a straight path from x to the detector.

If the structure of the matrix varies from sample to sample, then the path-integrated absorption coefficient will also show ensemble variation. Consequently, the point-source response functions will show ensemble variation. Let us consider an assay system with n detectors. The point-source response function for the i-th detector will be

$$f_i(x) = \frac{k}{d_i^2(x)} e^{-v_i(x)}$$

where $v_i(x)$ is the variable path-integrated absorption-coefficient function for the i-th detector and $d_i(x)$ is the distance from point x to the i-th detector. Let S represent the set of allowed point-source response functions. Also, let S be the convex hull of a fundamental set, S'. Let the fundamental set be

$$S' = \left\{ f : f = \left(\frac{k}{d_1^2(x)} e^{-v_1(x)}, \ldots, \frac{k}{d_n^2(x)} e^{-v_n(x)} \right); \right.$$
$$\left. \left(v_1(x), \ldots, v_n(x) \right) \in Q \right\}$$

where Q is an n-dimensional set of allowed path integrated absorption-coefficient functions.

Suppose the source material is uniformly distributed in the volume V of the sample. Then the complete response set for the zeroth moment is

$$\tilde{C}(0, u) \left\{ c : c = \frac{u}{V} \int f(x)\, dx \quad , \quad f \in S \right\}$$

Since S is convex, so is $\tilde{C}(0, u)$. Also, the class $\{\tilde{C}(0, u), u > 0\}$ has the property of proportionality. Thus the relative resolution is the expansion of $\tilde{C}(0, 1)$. So, according to Theorem 3,

$$z(0) = e\left(\tilde{C}(0, 1) \right)$$

$$= \underset{w \in W}{\text{minimum}} \ \underset{c,d \in \tilde{C}(0,1)}{\text{maximum}} \frac{\langle w, c \rangle}{\langle w, d \rangle}$$

where W is the set of real n-tuples for which

$$\langle w, c \rangle \langle w, d \rangle > 0$$

for all c and d in $\widetilde{C}(0, 1)$.

The calculation of $z(0)$ can be simplified somewhat. Since S is the convex hull of a fundamental set of response functions we can show that $\widetilde{C}(0, 1)$ is the convex hull of a fundamental set of responses, B. We shall prove subsequently that $\widetilde{C}(0, 1) = \text{ch}(B)$ where

$$B = \left\{ d : d = \frac{1}{V} \int f(x)\, dx \quad , \quad f \in S' \right\}$$

Thus the relative resolution can be found by applying the above min-max algorithm to the set B rather than to $\widetilde{C}(0, 1)$.

It is instructive to consider a numerical example. Let us consider a 2-detector assay of a sample whose domain X is the unit interval $[0, 1]$. The local absorption coefficient, $\mu(x)$, varies linearly through each sample as

$$\mu(x) = 2ax \quad , \quad a \in A$$

where A is a set of positive real numbers. Thus the local absorption coefficient increases from $x = 0$ to $x = 1$, and the rate of increase varies from sample to sample. The detectors are located along the axis of the sample, with detector 1 to the left and detector 2 to the right of the sample. Let the point-source response functions in the fundamental set S' be of the form

$$f_i(x) = e^{-v_i(x)} \quad , \quad i = 1 \text{ or } 2$$

where

$$v_1(x) = \int_0^x \mu(x)\, dx \quad , \quad v_2(x) = \int_x^1 \mu(x)\, dx$$

are the path-integrated absorption coefficient functions. Combining the above relations, the fundamental set of point-source response functions is

$$S' = \left\{ f : f = \left(e^{-ax^2}, e^{-a(1-x)^2} \right) \quad , \quad a \in A \right\}$$

The complete set of point-source response functions is $S = \text{ch}(S')$.

Let us suppose that the source material is uniformly distributed throughout the sample. Thus the zeroth-moment complete response set for u grams is

$$\widetilde{C}(0, u) = \left\{ c : c = u \int_0^1 f(x)\, dx \quad , \quad f \in S \right\}$$

$$= \text{ch} \left\{ c : c = u \int f(x)\, dx \quad , \quad f \in S' \right\}$$

$$= \text{ch}(B) \quad .$$

The elements of the fundamental response set B are

$$c_1 = \int f_1(x)\, dx = \frac{1}{2}\sqrt{\frac{\pi}{a}}\, \text{erf}(a)$$

$$c_2 = \int f_2(x)\, dx = \frac{D(a)}{\sqrt{a}}$$

where $\text{erf}(a)$ is the error function and $D(a)$ is Dawson's integral [25].

Since $\widetilde{C}(0, u)$ is the convex hull of B and the class $\{\widetilde{C}(0, u), u > 0\}$ has the property of proportionality, the relative resolution for the zeroth moment is

$$z(0) = \underset{w \in W}{\text{minimum}} \ \underset{c, d \in B}{\text{maximum}} \ \frac{\langle w, c \rangle}{\langle w, d \rangle}$$

where W is the set of real pairs such that

$$\langle w, c \rangle \langle w, d \rangle > 0$$

for all c and d in B.

Using the explicit expressions for c, the above min-max relation becomes

$$z(0) = \operatorname*{minimum}_{w \in W} \operatorname*{maximum}_{a,b \in A} \sqrt{\frac{b}{a} \frac{\frac{w_1 \sqrt{\pi}}{2} \operatorname{erf}(a) + w_2 D(a)}{\frac{w_1 \sqrt{\pi}}{2} \operatorname{erf}(b) + w_2 D(b)}} .$$

To obtain the relative mass resolution for detector 1 alone we set $w_2 = 0$ and solve this relation. Similarly if we hold $w_1 = 0$, we obtain the resolution for detector 2 alone. If both w_1 and w_2 are free to vary, we obtain the resolution of the 2-detector set-up.

As a numerical example, let A equal the set of real numbers between 1.2 and 1.4. Then the following results are obtained:

$$
\begin{aligned}
\text{detector 1 only:} \quad & z(0) = 1.032 \\
\text{detector 2 only:} \quad & z(0) = 1.200 \\
\text{both detectors :} \quad & z(0) = 1.0022
\end{aligned}
$$

This means that detector 1, when used alone, yields a relative mass resolution of 1.032, or an incremental resolution of the source mass of 3.2%. Likewise detector 2 alone has a relative resolution of 1.20 or an incremental resolution of 20%. The reason for this large difference in relative resolution is related to the difference in positioning of the detectors relative to the regions of strong absorption in the sample. Detector 1 (to the left of the sample) faces the low-absorption side and thus is more sensitive to the contents of the sample, while detector 2 (on the right) faces the region of high absorption coefficient. In light of the large difference between the two detectors when used separately, it is remarkable how much the resolution is improved when they are used in conjunction. The 2-detector system has a relative resolution of 1.0022, or an incremental resolution of only 0.22%.

These results can readily be demonstrated graphically since we have only 2 detectors. In figure 3.11.1 we plot the complete

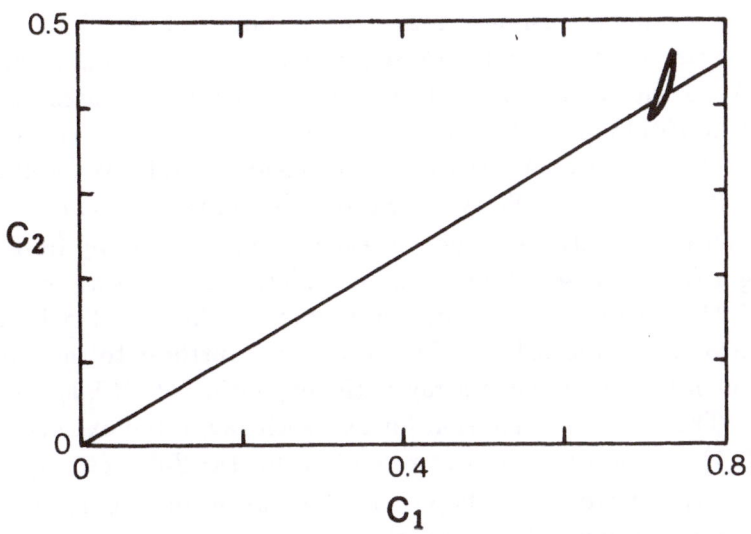

Figure 3.11.1 Complete response set for a two-detector assay system.

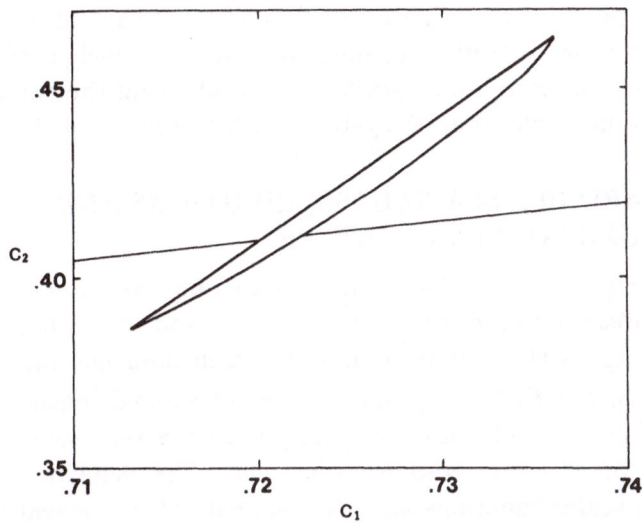

Figure 3.11.2 Expanded view of the complete response set shown in figure 3.11.1.

response set $\widetilde{C}(0,1)$, and in figure 3.11.2 we show an enlargement of the complete response set. The response of the first detector varies from 0.713 to 0.736 (for fixed mass of source material), so the relative resolution (when detector 1 is used alone) is $0.736/0.713 = 1.032$. Similarly the response of the second detector varies between 0.386 and 0.465, yielding a relative resolution of $0.465/0.386 = 1.200$. Now consider the convex region of paired responses for both detectors together, $\widetilde{C}(0,1)$. Any ray from the origin which passes through the complete response set, traverses $\widetilde{C}(0,1)$ over only a short distance as shown by the thin lines in figures 3.11.1 and 3.11.2. The ratio of the furthest to the closest point in $\widetilde{C}(0,1)$ on such a ray is the expansion of $\widetilde{C}(0,1)$ *on that ray*. The greatest expansion for any such ray is the expansion of $\widetilde{C}(0,1)$, or the relative mass resolution for the 2-detector system. It is evident from the shape and orientation of $\widetilde{C}(0,1)$ that its expansion is much less than the expansion of the response sets for either detector alone. In short, the 2-detector assay system is much better than either detector alone, and the concept of expansion provides a computationally feasible means of obtaining a physically meaningful assessment of just how much better. Furthermore, the min-max algorithm for evaluating the expansion is readily applicable to assay systems with any number of detectors.

3.12 VARIABLE SPATIAL DISTRIBUTIONS AND AUXILIARY PARAMETERS

In this section we shall consider ensemble variation of both the spatial distribution of the source material and of auxiliary parameters (such as the matrix structure) which influence the response functions. Let $\widetilde{R}(0,1)$ represent the set of allowed spatial distributions of one unit of source material, and let S represent the set of allowed point-source response functions. The elements of $\widetilde{R}(0,1)$ are real scalar functions and the elements of S are real n-vector functions. The complete response set for the zeroth moment is

$$\widetilde{C}(0,1) = \left\{ c : c = \int r(x) f(x) \, dx \quad , \quad r \in \widetilde{R}(0,1) \, , \, f \in S \right\}$$

The first point to note is that convexity of $\tilde{R}(0,1)$ and of S is not sufficient to assure the convexity of $\tilde{C}(0,1)$. To demonstrate this, let r and r' be elements of $\tilde{R}(0,1)$ and let f and f' be elements of S. In order for $\tilde{C}(0,1)$ to be convex, we must be able to find elements s in $\tilde{R}(1,0)$ and g in S such that, for any $0 \le a \le 1$,

$$a \int rf\, dx + (1-a) \int r'f'\, dx = \int sg\, dx \quad .$$

Since $\tilde{R}(0,1)$ may be simply ch $\{r, r'\}$ and S may be just ch $\{f, f'\}$, it is sufficient to ask: can we find numbers b and c in the interval $[0,1]$ such that

$$a \int rf\, dx + (1-a) \int r'f'\, dx = \int \big(br + (1-b)r'\big)\big(cf + (1-c)f'\big)dx$$

Since r and r' are scalar functions and f and f' are n-vector functions, this relation constitutes n independent algebraic equations in b and c. In general, no non-trivial solution need exist. We can, however, obtain a partial characterization of the sets $\tilde{R}(0,1)$ and S for which $\tilde{C}(0,1)$ is convex, for which we need the following definition.

DEFINITION 4 Let G be a set of real n-vector functions defined on the domain Y. If there is a real n-vector function q defined on Y and a scalar function p defined on a parameter set A, such that for any element g of G there is a value of a in A for which

$$g(y) = p(a)q(y) \qquad \text{for all} \qquad y \in Y$$

then the set G is *separable* in Y.

THEOREM 6 Let T be a convex set of real scalar functions on the domain X, and let G be a convex set of real n-vector functions on X. Suppose that either T or G is separable in x. Then the set:

$$C = \left\{ c : c = \int r(x)f(x)\, dx \quad , \quad r \in T\ ,\ f \in G \right\} \qquad (3.12.1)$$

is convex. Furthermore, if T and G are the convex hulls of sets T' and G' respectively, $T = \mathrm{ch}(T')$ and $G = \mathrm{ch}(G')$, then C is the convex hull of the following fundamental set C' defined by

$$C' = \left\{ c : c = \int r(x)f(x)\,dx \quad, \quad r \in T' \,, \; f \in G' \right\} \qquad (3.12.2)$$

PROOF (i) We shall begin by proving that C in eq.(1) is convex. Let us suppose that G is separable in x. For any r and s in T and any f and g in G, let

$$c = yrf + (1 - y)sg \qquad (3.12.3)$$

for $0 \leq y \leq 1$. Now

$$\int r(x)f(x)\,dx \qquad \text{and} \qquad \int s(x)g(x)\,dx$$

belong to C by definition. Thus C is convex if

$$\int c(x)\,dx$$

belongs to C. Since G is separable in x, there is a scalar function p defined on a parameter set A, and an n-vector function q defined on X such that f and g can be represented as

$$f(x) = p(a)q(x) \qquad \text{and} \qquad g(x) = p(b)q(x) \; . \qquad (3.12.4)$$

Thus let
$$h(x) = \big(yp(a) + (1 - y)p(b)\big)q(x) \qquad (3.12.5)$$

which belongs to G by convexity. Also define

$$u(x) = \frac{yp(a)}{yp(a) + (1 - y)p(b)}\, r(x) \quad + \quad \frac{(1 - y)p(b)}{yp(a) + (1 - y)p(b)}\, s(x)$$
$$(3.12.6)$$

which belongs to T by convexity. Now eqs.(3) to (6) may be combined to shown that

$$c = uh \quad .$$

Since u belongs to T and h belongs to G, we conclude that

$$\int c(x)\,dx = \int u(x)h(x)\,dx$$

belongs to C by definition. Thus C is convex. If T is separable rather than G a similar line of reasoning can be applied, and will not be elaborated.

(ii) Now we shall prove that $C = \text{ch}(C')$. It is evident that C' belongs to C, so it is sufficient to prove that C belongs to $\text{ch}(C')$. Let c be an arbitrary element of C, defined by

$$c = \int rf\,dx \quad \text{for} \quad r \in T \quad \text{and} \quad f \in G \quad .$$

Since T and G are the convex hulls of T' and G' respectively,

$$r = \sum y_i s_i \quad , \quad y_i \geq 0, \quad \sum y_i = 1, \quad s_i \in T'$$

$$f = \sum z_i g_i \quad , \quad z_i \geq 0, \quad \sum z_i = 1, \quad g_i \in G' \quad .$$

Thus

$$c = \int \sum_i y_i s_i(x) \sum_j z_j g_j(x)\,dx$$

$$= \sum_i \sum_j y_i z_j \int s_i(x) g_j(x)\,dx$$

which belongs to $\text{ch}(C')$. Thus C is a subset of $\text{ch}(C')$, and the proof is complete. QED

3.13 CONSTRAINED TIME-VARYING DISTRIBUTIONS

In all the examples which we have examined so far, the spatial distributions of the analyte material are stationary in time. In Chapter 1 we identified situations in which time-variation of the spatial distributions occurs. In this section we shall extend our design-analysis to include temporal variation.

As before, $\widetilde{R}$ represents the set of all allowed spatial distributions of the analyte material. These distributions $r(x,t)$ are continuous integrable functions on the spatial domain X of the sample, and on the time domain T of the measurement. $\widetilde{R}(h,u,t)$ represents the set of all allowed spatial distributions whose h-moment at time t equals u. $\widetilde{R}(h,u,t)$ is a set of time-varying functions whose h-moments may also vary in time. Thus, if $r(x,t)$ belongs to $\widetilde{R}(h,u,t)$, then

$$u = \int_X x[h]r(x,t)\,dx \quad .$$

However, the h-moment of r at some other time may differ from u.

The vector response function $f(x,t)$, which may also be a function of time, contains the point-source response functions of each of the n detectors. Measurements are to be performed during q time intervals, which we denote

$$I(j) = [t_j, t_j + \delta_j] \quad , \quad j = 1, 2, \ldots, q \tag{3.13.1}$$

These intervals may be contiguous in time, in which case δ_j becomes $t_{j+1} - t_j$. Alternatively, the intervals may be disjoint. In some cases the measurements may be instantaneous samples in time, in which case δ_j becomes an infinitesimal step in time, dt. The choice of the number and structure of the measurement intervals is a crucial element of the design process.

The instantaneous vector response to the spatial distribution $r(x,t)$ is

$$c(r, t) = \int\limits_{X} r(x, t) f(x, t)\, dx \quad \cdot$$

which is an n-vector. The time-averaged value of the measurement obtained during the j-th interval is

$$c(r, j) = \int\limits_{I(j)} \int\limits_{X} r(x, t) f(x, t)\, dx\, dt \qquad (3.13.2)$$

which is also an n-vector. The complete time-averaged value of the measurement vector obtained from the spatial distribution $r(x, t)$ is

$$c(r) = \big(c(r, 1), \ldots, c(r, q)\big) \qquad (3.13.3)$$

which is a vector of length nq. The complete response set for distributions whose h-moment at time t equals u is

$$\widetilde{C}(h, u, t) = \Big\{ c : c = c(r) \quad \text{for all} \quad r \in \widetilde{R}(h, u, t) \Big\} \qquad (3.13.4)$$

Theorem 4 of section 3.3 is a directly applicable criterion for the re-solvability of the values u and v (at time t) of the h-moment of the distribution. If $\widetilde{R}(h, u, t)$ is the convex hull of a set of fundamental distributions, then $\widetilde{C}(h, u, t)$ can be expressed as the convex hull of a fundamental response set, as in section 3.3.2. When such a decomposition is possible, it can be exploited to simplify the computation of the relative resolution. The statistical uncertainty of the measurement may be included precisely as outlined in section 3.3.3. Care must be taken however, to correctly account for any possible covariance of different measurements at different times.

In the following section we shall outline an application of these ideas.

3.14 EXAMPLE: FLOW RATE MEASUREMENT

In this section we shall briefly discuss the design of an assay system for flow-rate measurement. The sample is idealized as a

long one-dimensional pipe. Fluid is flowing in the pipe and car-
ries with it a radioactive tracer whose radiation is to be detected.
Measurements are performed along a section of pipe of length L.
There are N equally spaced detectors located at points $x_1, \ldots, x_N$
in the interval $X = [0, L]$. The detectors are highly collimated,
and the time-averaged response of the n-th detector to a source of
unit intensity is

$$f_n(x) = \delta(x - x_n) \qquad n = 1, 2, \ldots, N$$

where δ is Dirac's delta function.

A sequence of J measurements are performed successively during T seconds. Each measurement is of duration T/J seconds. The
j-th measurement interval is

$$I(j) = \left[\frac{j-1}{J}T, \frac{j}{J}T \right] \qquad j = 1, 2, \ldots, J$$

The concentration of tracer at any instant varies along the length
of the pipe. Since the tracer is carried along at the rate of the
flowing fluid, the spatial distribution varies in time. The spatial
distribution may vary in time due to dispersive phenomena as well.
The set of all allowed spatial distributions of the tracer, in a flow
regime whose flow rate is u, is represented by $\tilde{R}(u)$. This set is the
convex hull of a fundamental set $D(u)$. For example, one possible
choice of this fundamental set is

$$D(u) = \left\{ h : h(x, t) = g \exp\left(-g^2(x - ut)^2\right) \ , \quad g_1 \leq g \leq g_2 \right\} \ .$$

The design-analysis must assist us to make the following design
decisions.

1. How many measurements should be taken during the total
allotted time interval $[0, T]$, and how long should each measurement be. That is, what values should be chosen for J and for
T?

2. How many detectors should be employed along the measurement section $[0, L]$, and how should they be spaced? That is, what values should be chosen for N and for L?

Many additional design options are available if we wish to perform a more extensive analysis. One could readily consider the following possibilities.

3. Should the measurement intervals be of unequal duration?

4. Should the detector spacing be non-uniform along the measurement section?

5. Various degrees of collimation of the detectors may be investigated.

6. Numerous design concepts involving on-line adaptation of the measurement system may be explored. For example one may investigate adaptive positioning of the detectors or on-line determination of the duration of measurement of each detector. Consideration of adaptive assay is deferred to Chapter 6.

The response of the n-th detector in the j-th time interval to the spatial distribution $r(x, t)$ is

$$c_n(r, j) = \int_{I(j)} \int_X r(x, t) f_n(x) \, dx \, dt$$
$$= \int_{I(j)} r(x_n, t) \, dt \quad .$$

The vector response to the spatial distribution r is a vector of length NJ. Specifically

$$c(r) = \big(c_1(r, 1), \ldots, c_N(r, J)\big) \quad .$$

The complete response set for flow rate u is

$$\widetilde{C}(u) = \{c : c = c(r) \qquad \text{for all} \qquad r \in R(u)\} \quad .$$

Since $c(r)$ is linear in r and since $\widetilde{R}(u)$ is the convex hull of $D(u)$, we see that $\widetilde{C}(u)$ may be expressed as the convex hull of a fundamental response set $F(u)$ defined by

$$F(u) = \{b : b = c(r) \qquad \text{for all} \qquad r \in D(u)\}.$$

The complete response set $\widetilde{C}(u)$ is convex and compact, but the class of sets $\left\{\widetilde{C}(u), \ u > 0\right\}$ does not have the property of proportionality. Thus the criterion of distinguishability developed in section 3.3 must be applied. This may include the statistical uncertainty of measurement as described in section 3.3.3. The result of such an analysis is that, for any given value u of the flow rate, one establishes the complete range of flow-rate values v such that any allowed tracer distribution at flow rate v is distinguishable, to a specified level of statistical confidence, from any allowed tracer distribution at flow rate u.

NOTES

[1]   P. Kehler, Accuracy of Two-Phase Flow Measurement by Pulsed Neutron Activation Techniques, in *Multiphase Transport Fundamentals, Reactor Safety Applications*, Vol. 5, p.2483, Hemisphere Pub., 1980.

[2]   M. Perez-Griffo *et al*, Basic Two-Phase Flow Measurements Using N-16 Tagging techniques, NUREG/CR-0014, Vol. 2, p923, 1980.

[3] P. B. Barrett, An Examination of the Pulsed-Neutron Activation Technique for Fluid Flow Measurements, *Nucl. Eng. Design*, 74:183-92, (1982).

[4] P. A. M. Dirac, *The Principles of Quantum Mechanics*, Cambridge Univ. Press, 1958.

[5] Y. Ben-Haim, Convex Sets and Nondestructive Assay, *S. I. A. M. J. Alg. Disc. Methods*, accepted for publication.

[6] For sets in Euclidean space, compactness and closed-boundedness are equivalent. Compactness is however a much more general concept, whose properties we shall exploit.

[7] See ref. [7.2] of Chapter 2, p145.

[8] A. Friedman, *Foundations of Modern Analysis*, Dover 1982.

[9] See section 5.3 of ref. [7.2] of Chapter 2.

[10] M. H. Dickerson, K. T. Foster and R. H. Gudiksen, Experimental and Model Transport and Diffusion Studies in Complex Terrain, 29th Oholo Conf. on Boundary Layer Structure and Modelling, Zichron Ya'acov, Israel, March 1984.

[11] See refs. [7] and [12] of Chapter 1 and
R. E. Goans and G. G. Warner, Monte Carlo Simulation of Photon Transport in a Heterogeneous Phantom - I: Applications to Chest Counting of Pu and Am, *Health Physics*, 37: 533 - 42 (1979).

[12] See ref. [7] of Chapter 1.

[13] 1. C. D. Berger, R. E. Goans and R. T. Greene, The Whole Body Counting Facility at Oak Ridge National Laboratory: Systems and Procedure Review, ORNL/TM-7477 (1980).
The advantages of employing a high energy-resolution germanium detector are explored in
2. C. D. Berger and R. E. Goans, A comparison of the NaI-CsI Phoswich and a Hyperpure Germanium Detector Array for *In-Vivo* Detection of the Actinides, *Health Physics*, 40: 535-42 (1981).

[14] 1. J. D. Brain and P. A. Valberg, Deposition of Aerosol in The Respiratory Tract, *Amer. Rev. Respiratory Disease*, 120: 1325-73 (1979).
2. C. P. Yu and C. K. Diu, Total and Regional Deposition of Inhaled Aerosols in Humans, *J. Aerosol Sci.*, 14: 599-609 (1983).

[15] 1. J. D. Brain *et al*, Pulmonary Distribution of Particles Given by Intratracheal Instillation or by Aerosol Inhalation, *Environmental Research*, 11: 13-33 (1976).
2. S. M. Morsy *et al*, A Detector of Adjustable Response for the Study of Lung Clearance, *Health Physics*, 32: 243-51 (1977).

[16] This approach can be readily generalized to include a model of aerosol dispersion which includes the relative probability of aerosol deposition in different areas of the pulmonary region. Let p_i be the probability that a given particle will be deposited in the i-th sub-interval. The probability of a particular spatial distribution $(m_1, \ldots, m_N)$ is

$$p(m_1, \ldots, m_N) = \frac{M!}{m_1! \cdots m_N!} p_1^{m_1} \cdots p_N^{m_N} \quad .$$

[17] 1. I. S. Boyce, J. F. Cameron and D. Pipes, Proc. Symp. on Nuclear Techniques in the Basic Metal Industries, vol.1, p155, IAEA, 1973.
2. R. Bevan, T. Gozani, and E. Elias, Nuclear Assay of Coal, Electric Power Research Institute report EPRI-FP-989, vol.6, 1979.
3. E. Elias, W. Pieters and Z. Yom-Tov, Accuracy and Performance Analysis of a Nuclear Belt Weigher, Nucl. Instr. Meth., 178: 109-115 (1980).
4. J. B. Cummingham et al, Bulk Analysis of Sulfur, Lead, Zinc and Iron in Lead Sinter Feed Using Neutron Inelastic Scatter Gamma-Rays, Int. J. Appl. Rad. Isot., 35: 635-43 (1984).

[18] See refs. cited in ref. [16.1] of Chapter 1 and:
J. A. Oyedele, Spatial Effects in Radiation Diagnosis of Two-Phase Systems, Int. J. Appl. Rad. Isot., 35: 865-73 (1984).

[19] T. A. Boster, Source of Error in Foil Thickness Calibration by X-ray Transmission, J. Appl. Phys., 44: 3778-81 (1973).

[20] J. A. Oyedele, The Bias in On-Line Thickness Calibration by Radiation Transmission, Nucl. Instr. Meth., 217: 507-14 (1983).

[21] 1. H. Harmuth, Transmission of Information by Orthogonal Functions, Springer-Verlag, 1972.
2. S. Tzafestas and N. Chrysochoides, Nuclear Reactor Control Using Walsh Function Variational Synthesis, Nucl. Sci. Eng., 62: 763-70 (1977).

[22] Other choices might be

$$A = \left\{ a : (a_0 - T)^2 \le P \quad , \quad \sum_1^J a_j^2 \le Q \right\}$$

or

$$A = \{ a : T - D \le a_j \le T + D \quad , \quad j = 0, 1, 2, \ldots, J \}$$

[23] The major disadvantage of the Rademacher functions is that they do not form a complete set, since they are all (for $j > 0$) odd functions on $[0, 1)$ about the point $x = 1/2$. This can be remedied by employing the Walsh functions which do form a complete set. However much of the simplicity of our results would thereby be lost.

[24] Thorough expositions of dynamic programming may be found in many sources, including the following.
1. R. Bellman, *Dynamic Programming*, Princeton University Press, 1957.
2. R. Bellman, *Introduction to the Mathematical Theory of Control Processes*, Vol I, Academic Press, 1967.
3. R. Bellman, *Introduction to Matrix Analysis*, McGraw-Hill, 1970.
4. R. Bellman, *Methods of Nonlinear Analysis*, Academic Press, 1973.

[25] M. Abramowitz and I. Stegun, *Handbook of Mathematical Functions*, Dover, 1982.

CHAPTER 4

PROBABILISTIC INTERPRETATION
OF
MEASUREMENT

4.1 PROBABILITITY DENSITY OF THE MEASUREMENT

The previous two Chapters were devoted to developing an efficient computerizable algorithm for evaluating a deterministic measure of performance. We found that this criterion can be used to quantitatively compare alternative assay-system designs. By such comparison one can seek to optimize the number of detectors and their deployment around the sample as well as other related design parameters. However, the deterministic criterion makes no distinction between likely and unlikely spatial distributions of source material in the sample. As a consequence, we concluded that the deterministic measure of performance tends to give a conservative estimate of the relative mass resolution. Its utility is the speed and simplicity with which one may compare a large number of assay-system designs.

The basic concept which allows us to obtain a more realistic evaluation of the resolution capability of the assay system is the probability density of the measurement vector. This concept is fundamental to the probabilistic interpretation of measurement as well. In this Chapter we shall discuss probabilistic interpretation, and in so doing we shall develop several additional tools essential for the probabilistic design analysis to be developed in Chapter 5.

4.1.1 Single Measurement of a Single Source Particle.

Let us consider one unit of source material located at position x in the homogeneous matrix of a sample which contains no other source material. Measurement from position y outside the sample for a duration of one second would yield a certain number of counts whose average is represented by the point-source response function $f(x, y)$. In section 2.2 we introduced two basic properties of the response function, embodied in the assumptions of linearity in time and linearity in mass. On the basis of these assumptions the total response (number of counts) obtained at detector position y in duration t seconds from m grams of source material at position x in the sample is $tmf(x, y)$.

Suppose this response does not exceed the value g for all possible positions of the source particle. We may conclude that the conditional probability is unity of obtaining a measurement not exceeding g, given that a single particle is in the sample. We shall represent this situation by

$$P(g|1) = 1 \qquad (4.1.1)$$

If the source particle is randomly positioned in the sample with equal probability of occurring in any equal-volume regions of the sample, the particle is said to have a *uniform random distribution* in the sample. Now suppose that when the particle is located somewhere in a fraction u of the volume of the sample the resulting measurement is less than or equal to g, and when the particle is located in the remainder of the sample the measurement exceeds g. If the particle has uniform random distribution in the sample the conditional probability is u that the measurement will not exceed g. That is

$$P(g|1) = u \qquad (4.1.2)$$

We may formalize and generalize these considerations as follows. Define the characteristic function $H(x)$ as

$$H(x) = \begin{cases} 0, & \text{if } x < 0 \\ 1, & \text{if } x \geq 0 \end{cases} \qquad (4.1.3)$$

The conditional probability distribution for obtaining a measurement not exceeding g, given a single source particle with uniform random distribution in the sample, is

$$P(g|1) = \frac{1}{V} \int H\big(g - tmf(x,y)\big)\, dx \qquad (4.1.4)$$

where V is the volume of the sample, t is the duration of the measurement, m is the particle mass, y is the detector position and the integral is over the entire volume of the container.

In many assay applications, such as those discussed in Chapters 2 and 3, it is usually reasonable to assume that the deposits of source material have uniform random distribution in the matrix of the sample. However, in some cases this assumption is far from accurate. For instance, in the assay of radioactive aerosol particles in the lungs [1] the distribution of particles along the bronchii and bronchioles may be random but non-uniform: the probability of deep penetration is likely to be smaller than the probability of particle-deposition near the entrance to the lung. Various models may be developed to describe the probability of deposition as a function of depth of penetration [2]. Another situation in which non-uniform random distribution of the source material may occur is in randomly filled containers. A loosely-packed sample may display non-uniform random distribution along the height due to gravitational settling, or may display non-uniform radial distribution due to rotational jostling (the tea-leaf effect).

Eq.(4) can be readily generalized to accommodate non-uniform random distribution of the source particles in the sample. Let $q(x)\, dx$ be the probability that the source particle is located in the infinitestimal volume dx centered at point x in the sample. Then the conditional probability distribution for obtaining a measurement not exceeding g, given a single source particle in the sample, is

$$P(g|1) = \int H\big(g - tmf(x,y)\big)q(x)\, dx \qquad (4.1.5)$$

4.1.2 Single Measurement of Identical Source Particles

In order to calculate the conditional probability distribution
of the response to more than one source particle, we assume that
the particles are small and sparse enough so that they may be
considered to be independently positioned in the sample. That is,
$q(x)$ is the probability density for location of a given source particle
at position x, regardless of how many other particles are present
in the sample. Furthermore, we adopt the simplifying assumption
that all the source particles are of the same mass; in section 4.1.4
we will show how this assumption can be avoided.

We shall assume [3] the existence of the conditional probability
density of the probability distribution given in eqs.(4) and (5).
This conditional density is defined as

$$p(g|1) = \frac{dP(g|1)}{dg} \qquad (4.1.6)$$

Thus $p(g|1)\,dg$ is the probability of a single source particle yielding
a measurement of $g \pm dg$.

From the assumption of linearity in mass, a sample containing
two source particles of mass m at positions x_1 and x_2 will yield
the measurement $mf(x_1, y) + mf(x_2, y)$. We wish to construct
the probability density $p(g|2)$ of the measurement of these two
particles. Since the two particles are independently positioned in
the sample, the probability of one particle contributing the value
h to the total measurement is $p(h|1)dh$. The second particle must
contribute the value $g - h$, for which the probability density is
$p(g - h|1)$. Since the value of h is arbitrary, the probability density
for the total measurement is obtained by multiplying these two
quantities and integrating over h:

$$p(g|2) = \int p(g - h|1)p(h|1)\,dh \qquad (4.1.7)$$

This result is readily generalized to the case of n identical source
particles located at positions $x_1, \ldots, x_n$ and yielding the measurement $mf(x_1, y) + \cdots + mf(x_n, y)$. The probability density of the

sum of n independent random variables is the n-fold convolution of their common density [4]. Thus the conditional probability density for the response to n independently located identical source particles is

$$p(g|n) = \int p(g - h|n - 1)p(h|1)\, dh \qquad (4.1.8)$$

Likewise, the conditional probability distribution is

$$P(g|n) = \int P(g - h|n - 1)p(h|1)\, dh \qquad (4.1.9)$$

4.1.3 Multiple Measurements of Identical Source Particles

It is an elementary matter to generalize these results to the case of multiple measurements of identical source particles of mass m. The multiple measurements may be performed at different positions around the container, and may include measurements of different radiations. If there are r measurements, each measurement is identified by a subscript k, is located at position y_k, and has its own response function $f(x, y_k)$. The r measurements may be represented individually as $g_1, \ldots, g_r$ or collectively as the measurement vector g. Analogous to eq.(5), the conditional probability distribution for r measurements, given that one source particle is in the sample, is

$$P(g_1, \ldots, g_r|1) = \int \prod_{k=1}^{r} H\big(g_k - tmf(x, y_k)\big)q(x)\, dx \qquad (4.1.10)$$

The probability density is assumed to exist, and may be expressed by:

$$p(g_1, \ldots, g_r|1) = \frac{d^r P(g_1, \ldots, g_r|1)}{dg_1 \cdots dg_r} \qquad (4.1.11)$$

The conditional density and distribution for the response to n identical source particles can be expressed exactly as in eqs.(8) and (9), where g and h represent response vectors, and $g - h$ is their term-by-term difference.

4.1.4 Multiple Measurements of Nonidentical Source Particles

To complete our derivation of the probability distribution of the measurement vector conditioned by the number of particles, let us consider the possibility that the source particles may assume any of w discrete values of mass, $m_1, \ldots, m_w$. The conditional probability distribution for r measurements of a single particle of mass m_i may be evaluated from eq.(10), and is denoted $P_i(g|1)$ where g represents the vector response. The corresponding conditional density is $p_i(g|1)$. The conditional probability density and distribution for n source particles of mass m_i are obtained by n-fold convolutions of P_i and p_i, as in eqs.(8) and (9). These convolutions are denoted $p_i(g|n)$ and $P_i(g|n)$. Now let g^i represent the vector response from all the source particles of mass m_i. Let n_i represent the number of source particles of mass m_i, for $i = 1, 2, \ldots, w$. Thus the conditional probability density and distribution for the mass distribution $n_1, n_2, \ldots, n_w$ are:

$$p(g|n_1, \ldots, n_w) =$$

$$\int \cdots \int p_w \left(g - \sum_{j=1}^{w-1} g^j \middle| n_w \right) \prod_{i=1}^{w-1} p_i(g^i|n_i) \, dg^1 \cdots dg^{w-1}$$

$$(4.1.12)$$

$$P(g|n_1, \ldots, n_w) =$$

$$\int \cdots \int P_w \left(g - \sum_{j=1}^{w-1} g^j \middle| n_w \right) \prod_{i=1}^{w-1} p_i(g^i|n_i) \, dg^1 \cdots dg^{w-1}$$

$$(4.1.13)$$

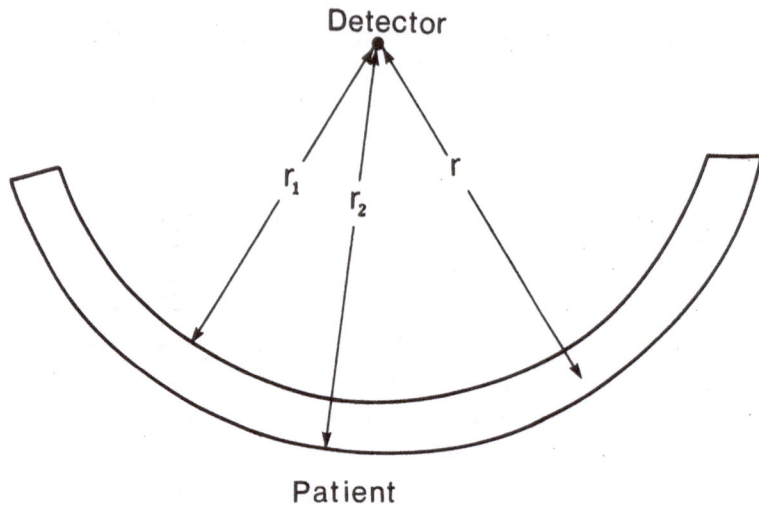

Detector

r_1 r_2 r

Patient

Figure 4.1.1 Idealized patient-detector geometry for whole-body assay.

4.1.5 Example: Medical Whole-Body Assay

We shall illustrate the calculation of the conditional probability density with a simplified example based on a standard whole-body assay technique. The patient is placed on a ∪-shaped bed facing upward. A single detector is located at the center of the circular arc defined by the longitudinal axis of the patient's body. The source-detector geometry may be idealized as in figure 4.1.1. The anterior and posterior surfaces of the "patient" are at distances r_1 and r_2 from the detector, and the source material may occur at any distance r between these two extremes. The response function of the detector is

$$f(r) = \frac{gspa}{4\pi r^2} e^{-\mu(r-r_1)} \qquad (4.1.14)$$

where the quantities g, s, p, a and μ are defined in connection with eq.(2.12.1). The distance r_1 is 60 cm.

An isotope commonly measured by this technique is the naturally occurring ^{40}K, which emits gamma radiation at 1.461 MeV. The linear absorption coefficient in soft tissue is 0.058 cm^{-1}. In

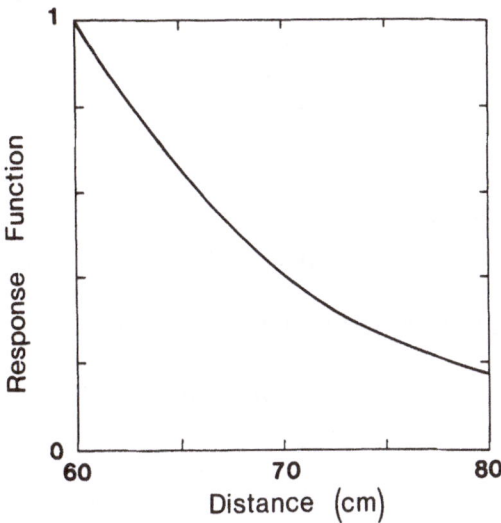

Figure 4.1.2 Normalized response function for idealized whole body assay.

figure 4.1.2 we show the response function, normalized to the response at the anterior surface of the patient. We note that the response is fairly linear up to a depth of about 10 cm. For simplicity we shall employ a linear approximation to the response function:

$$f(r) = -ar + b \qquad (4.1.15)$$

where a and b are positive quantities. Furthermore we shall assume that the probability of occurrence of a deposit of source material is the same at any depth in the sample. That is, the source material has a uniform random distribution over the interval $[r_1, r_2]$. We shall consider a measurement duration of 1 second.

These assumptions allow us to analytically evaluate the conditional probability distribution of the response to m grams of source material randomly distributed in identical deposits in the patient. From eq. (4) the conditional probability distribution given a single deposit is

$$P(g|1) = \frac{1}{r_2 - r_1} \int_{r_1}^{r_2} H\big(g - mf(r)\big)\, dr \qquad (4.1.16)$$

From the definition of the characteristic function H, the integrand equals unity only if

$$mf(r) \le g \qquad (4.1.17)$$

Employing eq.(15) we see that r must satisfy

$$r \ge \frac{mb - g}{ma} \qquad (4.1.18)$$

Since r is uniformly distributed on the interval $[r_1, r_2]$ we find the conditional probability distribution to be linear in g:

$$P(g|1) = \frac{mar_2 - mb + g}{ma(r_2 - r_1)} \qquad (4.1.19)$$

while the conditional probability density is constant:

$$p(g|1) = \frac{1}{ma(r_2 - r_1)} \qquad (4.1.20)$$

provided that the response is in the interval

$$m(b - ar_2) \le g \le m(b - ar_1) \qquad (4.1.21)$$

In other words, since a single source deposit has uniform random distribution, the response to a single source particle is also uniformly distributed.

Now let us consider the response to two source particles, each of mass $m/2$ grams. From eq.(7) the probability density is

$$p(g|2) = \int p(g - h|1)p(h|1)\, dh \qquad (4.1.22)$$

Employing eqs.(20) and (21) one finds the conditional density of the response to two identical source particles to be triangular:

$$p(g|2) = \begin{cases} \frac{4u^2(g - md_1)}{m^2} & ; \quad md_1 \le g \le \frac{1}{2}m(d_1 + d_2) \\ \frac{4u^2(md_2 - g)}{m^2} & ; \quad \frac{1}{2}m(d_1 + d_2) < g \le md_2 \end{cases}$$

where

$$u = \frac{1}{a(r_2 - r_1)}$$

$$d_1 = b - ar_2$$

$$d_2 = b - ar_1$$

This process can be continued indefinitely, to obtain the probability densities for greater subdivisions of the m grams of source material. In figure 4.1.3 we schematically show the conditional densities for 1, 2 and 4 identical source particles, where the total source mass is constant. For a single source particle the probability density is flat, and all allowable responses are equally likely. For finer subdivisions of the same source mass the probability density becomes more concentrated at intermediate values of the response, and the extreme response becomes increasingly less probable. We see from this example that the probabilistic analysis of the assay of spatially random material is very sensitive to the degree of subdivision of the source material. If the source mass is divided into many small particles, the probability of obtaining extreme values of the response becomes very low. This is in contrast to the deterministic measure of performance, which considers only the extreme responses, regardless of how unlikely they may be.

4.2 PROBABILITY DENSITY OF THE TOTAL SOURCE MASS

Our immediate interest in the conditional probability density of the response vector, $p(g|n)$, is in using it to evaluate the conditional probability density of the total contained source mass. That is, having measured a sample and obtained the response vector g,

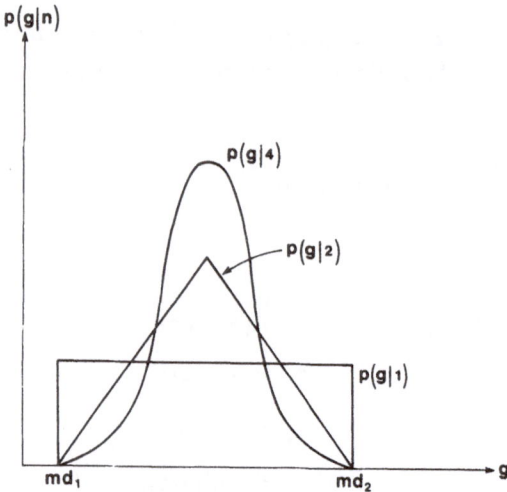

Figure 4.1.3 Schematic probability densities for the response to m grams of source material for subdivision of the source into 1, 2 or 4 particles.

we wish to know the probability density of the total source mass x in the sample, $p(x|g)$.

The starting point for construction of the desired probability density is knowledge of several basic quantities. First we must know the response functions $f(x, y_k)$ for each of the r measurements. Second, we must know the masses $m_1, \ldots, m_w$ of the source particles which may appear in the assayed sample. With this information we are able to evaluate the conditional probability $p(g|n) \, dg$ of obtaining a measurement vector g, given a distribution vector $n = (n_1, \ldots, n_w)$ of source-particle masses in the sample. Our aim is to calculate the conditional probability $p(x|g)dx$ that the source mass is $x \pm \frac{1}{2}dx$ given the measurement vector g. To achieve this we must know the marginal or *a priori* probability distribution $p(n)$ that a sample carries a distribution n of particle masses.

In the next section we shall use Bayes' rule to evaluate $p(x|g)$. In section 4.2.2 we shall discuss a plausible way of estimating $p(n)$ for some circumstances. In section 4.2.3 we shall discuss an alternative approach which does not require knowledge of $p(n)$. In section 4.2.4 we consider inclusion of statistical uncertainty of the measurement.

4.2.1 Bayes' Rule

We begin by developing an expression for the conditional probability density $p(n|g)$. From the definition of conditional probability one obtains

$$p(n|g) = \frac{p(n,g)}{p(g)} \tag{4.2.1}$$

where $p(n,g)$ is the multivariate probability density for n and g, and $p(g)$ is the marginal probability density for g. This marginal density may be expressed

$$p(g) = \sum_{n} p(g|n)p(n) \tag{4.2.2}$$

Again employing the definition of conditional probability, the numerator of eq.(1) may be written

$$p(n,g) = p(g|n)p(n) \tag{4.2.3}$$

Now eqs. (1)–(3) may be combined to yield

$$p(n|g) = \frac{p(g|n)p(n)}{\sum_{m} p(g|m)p(m)} \tag{4.2.4}$$

which is known as Bayes' rule [5]. Finally, the conditional probability $p(x|g)\,dx$ is evaluated as

$$p(x|g)\,dx = \sum_{n} p(n|g) \tag{4.2.5}$$

where the sum is over all values of the distribution vector n for which

$$x - \frac{1}{2}dx \leq \sum m_i n_i \leq x + \frac{1}{2}dx \tag{4.2.6}$$

4.2.2 The Poisson Distribution

Eqs.(4) and (5) show that evaluation of $p(x|g)$ requires knowledge of $p(g|n)$, whose derivation we have discussed at length in section 4.1, and knowledge of $p(n)$. In many situations it may be expected that the procedure by which the source material is loaded into the sample can be approximated as a Poisson process. The Poisson distribution describes the probability that any given number of independent random events will occur over a specified interval of space or time. A classic example of a Poisson process is the decay of a long-lived radioactive source. During a short interval of time (compared to the half life of the source) the independent random events are radioactive disintegrations, and the Poisson distribution describes the probability of any number of decays occurring in the interval. As an additional example, consider randomly filled containers of radioactive waste. If a batch of containers are filled from a large reservoir of waste, then the "random event" is a radioactive fragment falling in a particular container, which constitutes the "specified interval of space". When these conditions are satisfied and when the average number of particles per container is u, the probability that a given container holds n radioactive particles is

$$p(n) = \frac{e^{-u} u^n}{n!} \tag{4.2.7}$$

Likewise, when w different particle masses may be present and the average number of particles of mass m_i is u_i, the probability that the distribution of particle masses in a container is $n = (n_1, \ldots, n_w)$ is given by

$$p(n) = \prod_{i=1}^{w} \frac{e^{-u_i} u_i^{n_i}}{n_i!} \tag{4.2.8}$$

4.2.3 The Likelihood Function

The validity of our derivation of the conditional probability density $p(g|n)$ is an irrevocable consequence of the assumptions of linearity in time and mass of the response functions, and of assuming random and independent loading of the source particles. Unfortunately, further specification of the process by which the sample is filled is required to justify our use of the Poisson distribution to describe the probability that a container carries the distribution vector n of particle masses. If the sample-filling procedure does not approximate a Poisson process and if some other theoretical or empirical stochastic model cannot be found to describe the filling procedure, then we are entirely unable to evaluate the conditional probability $p(x|g)$.

However, in the spirit of maximum-likelihood decision theory [6] we might ask: what value of n is mostly likely to have given rise to the observed measurement? It is reasonable to answer that the most likely value of n is that which maximizes $p(g|n)$. This conditional probability may be evaluated as in section 4.1 and does not require knowledge of $p(n)$. Furthermore, it is reasonable to suppose that the relative likelihood for two values n and n' to have given rise to a specific measurement is related to the *likelihood ratio* $p(g|n)/p(g|n')$ [7]. In this way we are led to define the *likelihood function* for the number of source particles in the sample as

$$L(n|g) = \frac{p(g|n)}{\sum_m p(g|m)} \qquad (4.2.9)$$

This may be used as a substitute for the conditional probability distribution $p(n|g)$ when the latter is unattainable.

The likelihood function is based on a less complete stochastic model of the process by which the assayed containers are filled than is the conditional probability distribution. The immediate manifestation of this fact appears in comparison of eqs.(4) and (9): $p(n)$ is absent from the likelihood function. An important consequence is that for a sufficiently large number of particles, the conditional probability distribution diminishes with increasing number of particles more rapidly than the likelihood distribution. Upon dividing eq.(4) by eq.(9) one obtains

$$\frac{p(n|g)}{L(n|g)} = \frac{\sum\limits_{m} p(g|m)}{\sum\limits_{m} p(g|m)p(m)} p(n) \qquad (4.2.10)$$

Recall that $p(n)$ is a normalized distribution defined on the non-negative integer w-tuples. The normalization assures that $p(n)$ must approach zero for sufficiently large number of particles. The physical meaning of $p(n)$ would also imply this. Thus eq.(10) shows that there is a number q such that

$$p(n|g) < L(n|g) \qquad (4.2.11)$$

for all w-tuples for which

$$\sum_{1}^{w} n_i \geq q$$

Loosely speaking, this means that for a large number of particles, the tail of $L(n|g)$ is "larger" than the tail of $p(n|g)$. The additional information contained in the conditional probability distribution serves to decrease the estimate of the probability of large values of contained source mass. This is the primary incentive for using the conditional probability rather than the likelihood distribution. We should note however that eq.(11) does not imply that the variance of $p(n|g)$ is less than the variance of $L(n|g)$ [8].

4.2.4 Statistical Uncertainty

The function $p(x|g)\,dx$ expresses the probability that the contained source mass is $x \pm \frac{1}{2}dx$, given the measured response vector g. In many practical circumstances the empirical determination of the response vector is accompanied by statistical uncertainty. This uncertainty can readily be included in the estimate of the contained source mass. Let $v(h|g)$ be the conditional probability density that the true response vector (uncorrupted by statistical uncertainty) is h, given the measured value g. Then the conditional density of the contained mass may be estimated as

$$p'(x|g) = \int p(x|h)v(h|g)\,dh \qquad (4.2.12)$$

4.2.5 Example: Pu Assay With One Detector

In this section we shall develop a probabilistic interpretaton of the assay of low-level plutonium-containing waste. We shall consider a one-detector assay of cylindrical waste containers. The details of the sample and detector are presented in section 2.6. In section 2.12 we discussed the design of this assay system based on the deterministic measure of performance. We shall analyze the behavior of a single midplane detector located 100 cm from the sample center.

The waste-samples whose assay we must interpret contain variable numbers of identical 1-gram plutonium deposits. The procedure by which the plutonium particles are loaded into the sample is well represented by a Poisson process, and the average number of particles per sample is 10. Each sample is measured for 10 seconds. Our ultimate aim is to evaluate the conditional probability $p(n|g)$ that a sample contains n plutonium particles given the measured response g counts in 10 seconds. To achieve this goal we must proceed through the following stages:

1. Calculation of the conditional probability distribution $P(g|1)$ of the measured response, given a single source particle in the sample. This requires an integration on the volume of the sample, as indicated by eq.(4.1.4).

2. Calculation of the conditional probability density $p(g|1)$ of the measured response, given a single source particle in the sample. This calls for derivation of $P(g|1)$ with respect to g, as shown in eq.(4.1.6).

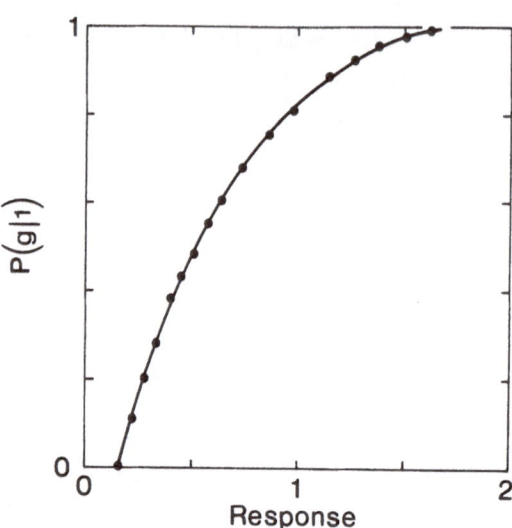

Figure 4.2.1 The conditional probability distribution of the response given one source particle in the sample, $P(g|1)$, versus the response, g. The source particle is one gram of ^{239}Pu. The measurement duration is 10 seconds. One detector on the sample midplane located 100 cm from the sample center.

3. Calculation of the conditional probability density $p(g|n)$ of the measured response, given n source particles in the sample. This requires the n-fold convolution of $p(g|1)$, as expressed iteratively by eq.(4.1.8).

4. Evaluation of the *a priori* probability of n source particles occurring in the sample. In our example this is represented by the Poisson distribution given in eq.(4.2.7).

5. Evaluation of the conditional probability $p(n|g)$ of n particles occurring in the sample, given the measured response g. This calls for application of Bayes' rule, eq.(4.2.4).

6. We shall compare the conditional probability distribution $p(n|g)$ with the likelihood function $L(n|g)$. The likelihood function is evaluated from $p(g|n)$ without requiring knowledge of $p(n)$, as shown in eq.(4.2.9).

In figure 4.2.1 we show the results of numerical evaluation of $P(g|1)$ (solid points). This is necessarily a monotonically increasing function of g, and can be accurately fit with a low-order polynomial as shown by the curve. The analytical derivative of this polynomial gives a good estimate of the probability density $p(g|1)$. This method of evaluating the probability density is more accurate than evaluating $p(g|1)$ directly from the numerical evaluation of $P(g|1)$ by finite difference. Small irregularities arise in $P(g|1)$ due to the finite step-size of the integration as shown by the points in figure 4.2.1. These irregularities are greatly magnified in $p(g|1)$ when it is evaluated from $P(g|1)$ by finite difference. Evaluating $p(g|1)$ from the polynomial fit to $P(g|1)$ eliminates this undesirable grid effect.

Since $p(g|1)$ is obtained as the derivative of a polynomial, its polynomial representation is also known. This polynomial can be numerically convoluted with itself to yield the value of $p(g|2)$ for a sequence of values of g. This result can in turn be fitted with a polynomial and convoluted numerically with $p(g|1)$ to produced values of $p(g|3)$ for a sequence of values of g. In this way $p(g|n)$ may be generated for as many successive values of n as needed.

In principle a polynomial representation of $p(g|n)$ can be evaluated analytically from the polynomial representation of $p(g|1)$. However, the order of the polynomial obtained in this way increases with n, and this results in polynomials of unnecessarily large order. Consequently the above numerical method is simpler if one wishes to generate distributions $p(g|n)$ for more than just a few values of n.

In figure 4.2.2 we show the conditional probability density of the measurement, $p(g|n)$, versus g for a range of values n. As the number of particles increases, the conditional probability of the response broadens and moves to higher values of g. In figure 4.2.3 we show (on the left ordinate) the average response, $\langle g \rangle$, versus the number of particles. Though this is a discrete function, we have represented it as a smooth curve for clarity. The average response increases linearly with the number of particles, as is to be expected from the assumption of linearity in mass and from the uniform random distribution of source particles in the sample. The

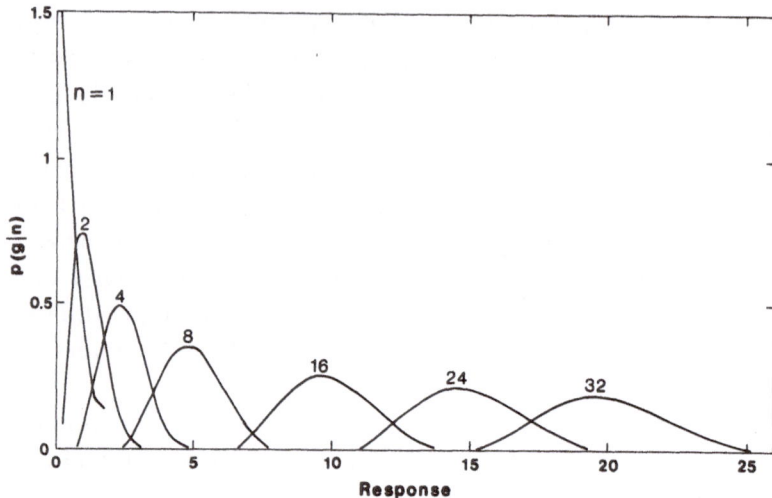

Figure 4.2.2 The conditional probability density of the response given n plutonium source particles, $p(g|n)$, versus the response, g.

thin lines show the average response plus and minus one standard deviation. On the right ordinate of figure 4.2.3 we show the relative error, $\sigma(g)/\langle g \rangle$, versus n. This is a steadily decreasing function of n, showing rapid decrease for small n, and gradual decrease for large n. As the number of source particles increases the probability decreases for occurrence of those spatial distributions of the source material which give rise to response deviating greatly from the mean.

In figure 4.2.4 we plot the Poisson distribution, $p(n)$ of eq.(7), with a mean value of 10. This is a nearly symmetric distribution, whose tail on the high-n side rapidly vanishes for $n > 22$. The standard deviation is approximately 3.2.

It is now an elementary matter to combine the calculations of the conditional probability density of the response, $p(g|n)$, with the marginal probability distribution of the number of particles $p(n)$, in order to evaluate the conditional probability distribution of the number of particles, $p(n|g)$. These distributions are shown in figure 4.2.5 versus n for various values of the response g. As

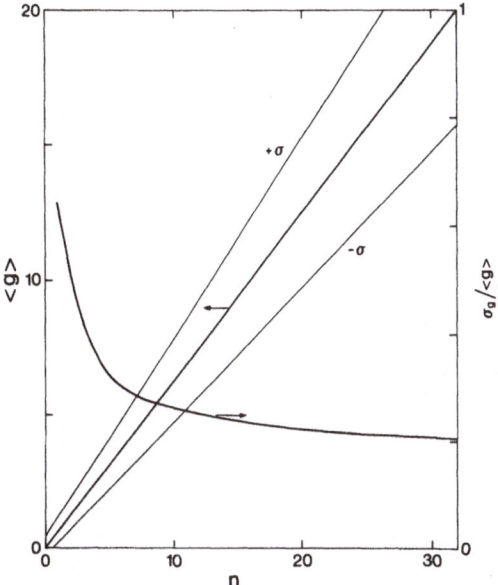

Figure 4.2.3 The left ordinate shows the average response $\langle g \rangle$ versus the number of plutonium particles. The thin lines show the average response plus and minus one standard deviation. The right ordinate shows the relative error of the response, $\sigma(g)/\langle g \rangle$.

is to be expected, $p(n|g)$ broadens and moves to greater values of n as the response increases. Curve P in figure 4.2.6 shows the average number of particles, $\langle n \rangle$, versus the response. This is an increasing function, though it is not strictly linear. This curve may be viewed as an average calibration of the system. Curve P in figure 4.2.7 shows the relative error $\sigma_n/\langle n \rangle$ versus the response, which is a decreasing function. Over most of the range of g, the relative error of the number of source particles takes a value not exceeding 0.2. This corresponds to a relative mass resolution of 1.2. This roughly means that the measurement of n source particles is distinguishable, at a confidence of one standard deviation, from the measurement of $1.2n$ source particles. This is to be compared with the deterministic relative mass resolution without statistical uncertainty of 12.8. We see clearly in this example that the

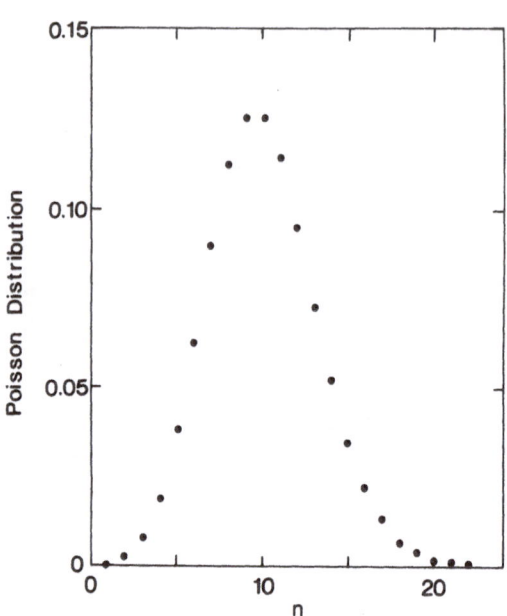

Figure 4.2.4 The marginal distribution of the number of plutonium particles per sample, $p(n)$, versus n.

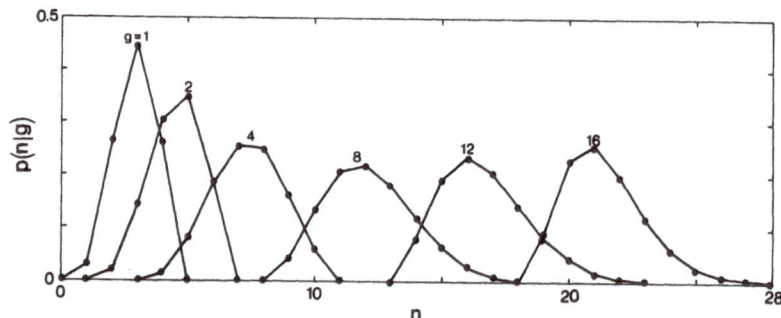

Figure 4.2.5 The conditional probability density $p(n|g)$ of the number of plutonium particles given the response g, versus the number of particles.

deterministic measure of performance is a very conservative estimate of the resolution capability of the assay system. Since the deterministic criterion does not account for the probability of different spatial distributions of source material, it yields a very large

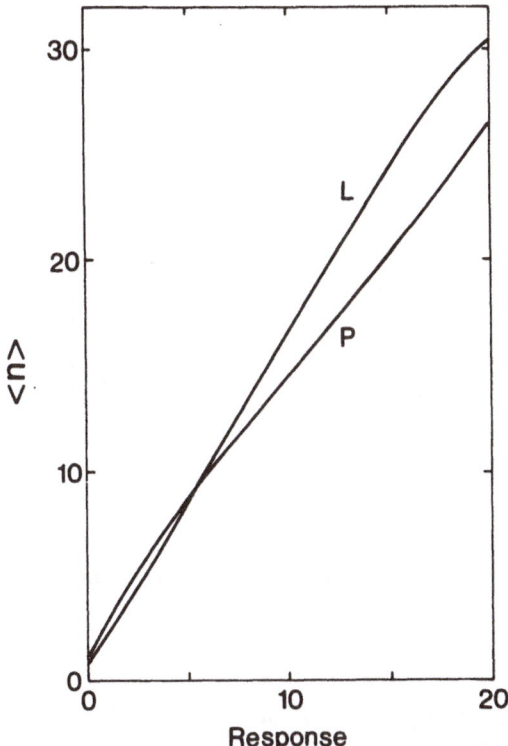

Figure 4.2.6 The average number of plutonium particles versus the response, based on the conditional probability density $p(n|g)$ and based on the likelihood function $L(n|g)$.

estimate of the relative mass resolution. The utility of the deterministic criterion is, as we have stressed, its ease of application to a large number of proposed assay-system designs. The deterministic criterion provides a useful tool for first-stage design analysis, by enabling quantitative comparison of alternative proposed designs. As we see in this example, though, the deterministic performance criterion does not necessarily provide a reliable absolute evaluation of the resolution capability of the assay system.

It is interesting to compare the likelihood function $L(n|g)$ with the conditional probability distribution $p(n|g)$. In figure 4.2.8 we plot $L(n|g)$ versus n for various values of the response. In compar-

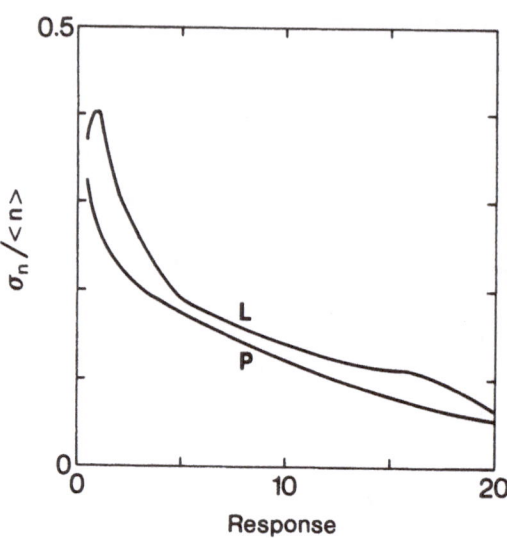

Figure 4.2.7 The relative error of number of plutonium particles, $\sigma_n/\langle n \rangle$, versus the response, for the conditional probability density $p(n|g)$ and for the likelihood function $L(n|g)$.

ing this figure with figure 4.2.5 the most notable difference is that $L(n|g)$ is generally a broader distribution than the corresponding $p(n|g)$. Referring to figure 4.2.6 we see that the mean of $L(n|g)$ (curve L) is less than the mean of $p(n|g)$ (curve P) for $g \leq 6$, and greater for $g > 6$. (Since $p(n)$ is not a non-increasing sequence, the result proven in note [8] does not hold, as this comparison shows). At very low response ($g < 0.7$), the variance of $L(n|g)$ is also less than the variance of $p(n|g)$. For greater response the variance of $L(n|g)$ exceeds that of $p(n|g)$. As shown in figure 4.2.7, the relative error of n is somewhat greater for $L(n|g)$ (curve L) than for $p(n|g)$ (curve P) over the range of g considered.

4.2.6 Example: Assay of Aerosol Particles in the Lungs

In the example discussed in the previous section the source particles were distributed in the sample with uniform random distribution. That is, a source particle has equal probability of occur-

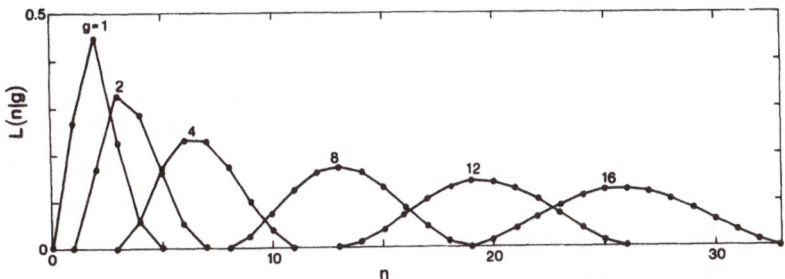

Figure 4.2.8 The likelihood function $L(n|g)$ of the number of plutonium particles versus the number of particles.

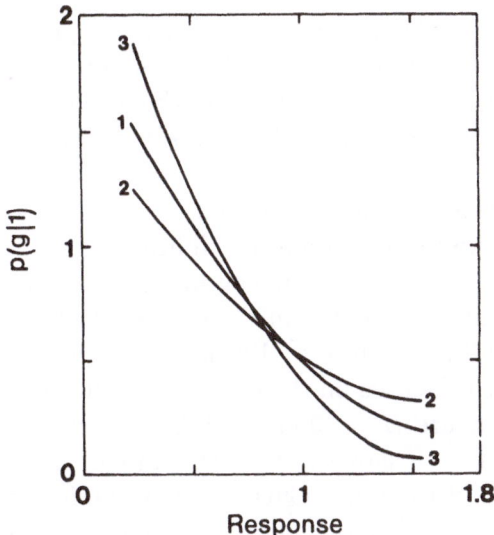

Figure 4.2.9 The conditional probability density of the measurement given one plutonium particle in the lungs.

ring in any equal-volume regions of the sample. In this section we shall illustrate the calculation of the conditional probability density of the measurement, $p(g|1)$, for a situation in which the spatial distribution of the source particle is random but not uniform. We have referred to this situation in connection with eq.(4.1.5).

Let us consider a single small particle containing ^{239}Pu, which

has been inhaled and deposited in the lungs. The point at which
the particle will be deposited is random, but the probability of
deposition is not uniform over the entire lung due to a number
of factors. For simplicity let us adopt a cylindrical model of the
lung, H cm high and R cm in radius. Let the coordinate origin
be at the center of the cylinder. Let $q(r, \theta, z)$ represent the probability density for deposition at point (r, θ, z) in the lung. Let this
function be represented by

$$q(r, \theta, z) = be^{-v_z(H-z)} e^{v_r(R-r)} \tag{4.2.13}$$

where b is a normalization constant and an implicit function of v_z
and v_r. If v_z is positive, the probability of deposition is a decreasing function of the penetration along the z axis. If v_r is positive,
the particle is more likely to be deposited near the center than
near the periphery. If the signs of v_z or v_r are reversed, the opposite distributions result. If v_z and v_r are both zero the source
particle has a uniform random distribution.

In figure 4.2.9 we show the conditional probability density of
the response given a single source particle in the lung. This calculation is for 0.01 gram of ^{239}Pu, measured for 1000 seconds. A
single detector is located on the midplane at 100 cm from the lung.
The dimensions are $H = 30$ cm and $R = 10$ cm. Three cases are
compared. Curve 1 shows a uniform random distribution of the
source particle in the lung. Curve 2 shows preferential deposition
at the top and around the periphery. Curve 3 shows preferential deposition at the top and along the central axis. As in the
previous example, the calculation of $p(g|1)$ forms the basis of the
probabilistic interpretation of the measurement, as expressed by
$p(n|g)$ and its moments. We see that the interpretation is likely to
be sensitive to the nature of the random distribution of the source
particles in the lung. Furthermore, this example illustrates that,
when detailed information is available on the spatial probability
distribution $q(r, \theta, z)$, this information can be readily incorporated
into the probabilistic analysis of the measurement.

4.3 BAYES' DECISION THEORY

4.3.1 General Formulation

In the previous sections we have developed expressions for two quantities of fundamental importance to the probabilistic interpretation of the measurement of spatially random material. The first is the conditional probability density of the response given the number and masses of source particles in the sample, $p(g|n)$. The second is the conditional probability density of the total source mass in the sample given a value for the response, $p(x|g)$. This latter probability density provides a direct and detailed probabilistic interpretation of a given measurement. The mean, $\langle n(g) \rangle$, may serve as a probabilistic calibration curve, while the standard deviation gives a succinct assessment of the uncertainty of the calibration which arises from spatial randomness of the source material.

One is likely to feel that the utility as well as the physical meaning of $\langle n(g) \rangle$ as a probabilistic calibration is not open to question. However, a number of other prescriptions for interpretation of a given measurement may seem no less reasonable, and yet lead to very different results. For example, one may wish to interpret the response g as having arisen from the most probable quantity of mass.

In this section we shall employ Bayes' decision theory to develop several different probabilistic interpretations. For clarity we shall consider the case of identical source particles throughout most of this section. This allows us to deal with the discrete probability distribution $p(n|g)$ rather than the continuous density $p(x|g)$. Generalization to non-identical source particles and to the continuous case do not involve fundamentally different considerations. We shall briefly discuss the estimation of a continuous parameter in section 4.3.7. Thorough treatments of this subject are not lacking [9].

The starting point for Bayes' decision theory is the choice of a cost function, C_{mn}. This represents the cost or penalty associated with deciding that a given measured sample contains m source particles when, in fact, it contains n particles. Our aim is to develop a rule for deciding on the number of source particles, in such a way that the average cost incurred will be a minimum.

We may develop an expression for the average cost as follows. Let G represent the set of all possible responses. Our decision-rule specifies the subset, G_m, of measurements which are to be interpreted as arising from m source particles, for $m = 0, 1, 2, \ldots$. Now suppose that there are in fact n source particles in a given sample. The cost of deciding that there are m particles is C_{mn}, and the probability of incurring this cost is just the probability p_{mn} of n particles giving rise to a response in the set G_m:

$$p_{mn} = \int_{G_m} p(g|n) \, dg \qquad (4.3.1)$$

Thus the average cost, given n particles in the sample, is

$$c_n = \sum_m C_{mn} p_{mn} \qquad (4.3.2)$$

Now, the probability of n source particles being in the sample is $p(n)$. Thus the overall average cost, or risk, is

$$R = \sum_n p(n) c_n \qquad (4.3.3)$$

Combining eqs.(1) to (3) yields the following expression for the risk:

$$R = \sum_n p(n) \sum_m C_{mn} \int_{G_m} p(g|n) \, dg \qquad (4.3.4)$$

Our decision rule must choose the sets G_m so as to minimize R.

We may express eq.(4) somewhat differently by exploiting the fact that the sets G_m are disjoint and that their union equals G, the set of all responses. Eq.(4) can be written

$$R = \sum_n p(n) C_{nn} \int_{G_n} p(g|n)\, dg$$

$$+ \sum_n p(n) \sum_{m \neq n} C_{mn} \int_{G_m} p(g|n)\, dg \qquad (4.3.5)$$

By employing the relation

$$G_n = G - \cup_{m \neq n} G_m \qquad (4.3.6)$$

the first sum in eq.(5) may be written as

$$\sum_n p(n) C_{nn} \int_{G_n} p(g|n)\, dg$$

$$= \sum_n p(n) C_{nn} \int_G p(g|n)\, dg$$

$$- \sum_n p(n) C_{nn} \sum_{m \neq n} \int_{G_m} p(g|n)\, dg \qquad (4.3.7)$$

Combining eqs.(5) and (7) and reversing the order of the second sum yields

$$R = \sum_n p(n) C_{nn}$$

$$+ \sum_m \int_{G_m} \sum_n p(n)\, (C_{mn} - C_{nn})\, p(g|n)\, dg \qquad (4.3.8)$$

(The sums on m and n may be allowed to cover all values because the summand vanishes for $m = n$). This formulation of the risk shows that only the differential cost, $C_{mn} - C_{nn}$, affects the choice of the decision regions G_m.

We may formulate the choice of the decision regions as follows. The risk R may be minimized by assigning each value, g, of the response to the decision region for which the corresponding integrand in eq.(8) is a minimum. That is, let

$$H_m(g) \equiv \sum_n p(n) \, (C_{mn} - C_{nn}) \, p(g|n) \qquad (4.3.9)$$

Then

$$g \in G_m \qquad (4.3.10)$$

if

$$H_m(g) < H_k(g) \qquad \text{for all} \qquad k \neq m \qquad (4.3.11)$$

If $H_m(g) = H_k(g)$ for some k, it does not matter to which of these decision regions we assign g.

For any given value of g, the comparison in eq.(11) is unaffected if we divide both sides by any positive quantity $D(g)$. Let us choose

$$D(g) \equiv \sum_n p(n) p(g|n) \qquad (4.3.12)$$

Now define

$$I_m(g) = \frac{H_m(g)}{D(g)}$$
$$= \sum_n (C_{mn} - C_{nn}) \, p(n|g) \qquad (4.3.13)$$

The decision rule is now

$$g \in G_m \qquad (4.3.14)$$

if

$$\sum_n (C_{mn} - C_{nn}) \, p(n|g) < \sum_n (C_{kn} - C_{nn}) \, p(n|g) \qquad (4.3.15)$$

for all $k \neq m$. Eq.(15) states that the decision regions are determined by the differential cost averaged on the distribution $p(n|g)$.

4.3.2 Binary Decisions

In some applications we are only interested in deciding whether or not there are at least N source particles in the sample. Thus we need only consider four costs C_{mn}, where $m = 0$ represents the decision that there are not more than N particles, $m = 1$ represents the opposite decision, $n = 0$ means that there are in fact no more than N source particles in the sample, and $n = 1$ means that the opposite situation prevails.

Let us denote by p_0 the probability that not more than N source particles occur in the sample, and let p_1 denote the opposite case. Thus

$$p_0 = \sum_{n \leq N} p(n) \qquad (4.3.16)$$

$$p_1 = \sum_{n > N} p(n) \qquad (4.3.17)$$

Similarly, let $p_0(g)$ and $p_1(g)$ be the conditional probabilities of g, given not more than N, and strictly more than N, particles in the sample. Employing the definition of conditional probability one can readily establish the following relations:

$$p_0(g) = \frac{\sum_{n \leq N} p(g|n)p(n)}{p_0} \qquad (4.3.18)$$

$$p_1(g) = \frac{\sum_{n > N} p(g|n)p(n)}{p_1} \qquad (4.3.19)$$

Now the risk may be adapted from eq.(8) as

$$R = \sum_{n} p_n C_{nn} + \sum_{m} \int_{G_m} \sum_{n} p_n \left(C_{mn} - C_{nn} \right) p_n(g) \, dg \qquad (4.3.20)$$

From this equation we may readily construct the decision regions as:

$$g \in G_0 \qquad (4.3.21)$$

if

$$p_1 \left(C_{01} - C_{11} \right) p_1(g) < p_0 \left(C_{10} - C_{00} \right) p_0(g) \qquad (4.3.22)$$

and

$$g \in G_1 \qquad (4.3.23)$$

if the opposite inequality holds in relation (22).

4.3.3 Minimum Probability of Error

The choice of a cost function C_{mn} is largely a subjective matter. In this and the next two sections we shall consider three cost functions with different physical meanings and which lead, in some cases, to quite different decision rules.

Consider the cost function

$$C_{mn} = \begin{cases} 1 , & \text{if } m \neq n \\ 0 , & \text{if } m = n \end{cases} \qquad (4.3.24)$$

That is, no cost is incurred by a correct decision, and a unit cost is incurred by an incorrect decision. Applying this cost function to the risk, eq.(4), yields

$$R = \sum_n p(n) \sum_{m \neq n} \int_{G_m} p(g|n) \, dg \qquad (4.3.25)$$

The inner sum (on m) is the probability of making an incorrect decision, given n particles in the sample. The outer sum averages this on n, to give the overall probability of error. Since the decision regions G_m are chosen to minimize the risk, R becomes the minimum probability of error.

This cost function can be interpreted in a different and equally interesting manner. By substituting the cost function into eq.(8) the risk becomes

$$R = \sum_m \int_{G_m} \sum_{n \neq m} p(n)p(g|n) \, dg \qquad (4.3.26)$$

The integrand $I_m(g)$ defined in eq.(13) becomes

$$I_m(g) = \sum_{n \neq m} p(n|g) \tag{4.3.27}$$

Thus the decision rule is

$$g \in G_m \tag{4.3.28}$$

if

$$\sum_{n \neq m} p(n|g) < \sum_{n \neq k} p(n|g) \tag{4.3.29}$$

for all $k \neq m$. Now it is evident that ineq.(29) is equivalent to

$$p(m|g) > p(k|g) \qquad \text{for all} \qquad k \neq m \tag{4.3.30}$$

In other words, the cost function eq.(24) generates a decision rule which selects the most probable value of m.

4.3.4 Quadratic Penalty

An alternative attractive cost function assigns a penalty which increases with the square of the error:

$$C_{mn} = (m - n)^2 \tag{4.3.31}$$

Substituting this cost function into eq.(8) yields the following expression for the risk:

$$R = \sum_m \int_{G_m} \sum_n (m - n)^2 p(n) p(g|n) \, dg \tag{4.3.32}$$

Referring again to eq.(13), $I_m(g)$ becomes

$$I_m(g) = \sum_n (m - n)^2 p(n|g) \tag{4.3.33}$$

Consequently, the decision rule is

$$g \in G_m \qquad (4.3.34)$$

if

$$\sum_n (m-n)^2 p(n|g) < \sum_n (k-n)^2 p(n|g) \qquad (4.3.35)$$

for all $k \neq n$.

We will now show that this decision rule leads to a very simple and physically meaningful result. Eq.(33) shows that $I_m(g)$ is a quadratic function of m. Let us for a moment allow m to be a continuous variable. Thus for constant g, $I_m(g)$ achieves a minimum at

$$m = \sum_n n p(n|g) = \langle n(g) \rangle \qquad (4.3.36)$$

This shows that a measured response, g, is to be interpreted as having arisen from the number of particles closest to the average, $\langle n(g) \rangle$. In other words, the quadratic penalty function generates a decision rule which is essentially the average calibration.

The mean and the maximum coincide when the probability distribution $p(n|g)$ is symmetric in n. In other words, the minimum-probability-of-error and the quadratic-penalty cost functions generate the same decison rules when $p(n|g)$ is symmetric. However if $p(n|g)$ is not symmetric in n, these decision rules may be greatly different. A simple hypothetical example will demonstrate this point. Let $p(n|g)$ be represented by the exponential distribution:

$$p(n|g) = b(g)e^{-cgn} \qquad (4.3.37)$$

where c is positive and $b(g)$ is a normalization constant. For given g, the mean $\langle m(g) \rangle$ and most probable $\hat{m}(g)$ values are

$$\langle m(g) \rangle = cg \qquad (4.3.38)$$
$$\hat{m}(g) = 0 \qquad (4.3.39)$$

Thus the quadratic penalty generates a linear decision rule, while the minimum-probability-of-error cost function causes every response to be interpreted as $m = 0$.

4.3.5 Biased Quadratic Penalty

As a final example of a cost function consider

$$C_{mn} = (m - n - x)^2 \qquad (4.3.40)$$

where x is any real number. Let us compare the cost incurred by under-estimation and over-estimation by the positive magnitude k:

$$C_{n-k,n} = (-x - k)^2 \qquad (4.3.41)$$
$$C_{n+k,n} = (-x + k)^2 \qquad (4.3.42)$$

Thus, if x is positive, an under-estimate is penalized more than an over-estimate of the same magnitude. This is a useful cost function in situations where one wants a certain degree of conservatism in the decision rule.

The risk resulting from this cost function is

$$R = \sum_m \int_{G_m} \sum_n (m - n - x)^2 p(n) p(g|n) \, dg \qquad (4.3.43)$$

The decision rule is based on comparison of

$$I_m(g) = \sum_n (m - n - x)^2 p(n|g) \qquad (4.3.44)$$

As in the case of the quadratic penalty, $I_m(g)$ achieves its minimum at

$$m = \langle n(g) \rangle + x \qquad (4.3.45)$$

Thus, the biased quadratic cost function achieves a conservative decision-rule (for $x > 0$) by linearly shifting the calibration curve.

4.3.6 Example: Plutonium Assay With One Detector

This example will consider various Bayes decision schemes for the one-detector plutonium assay problem discussed in section 4.2.5.

First we consider the quadratic-cost decision rule. For a given value, g, of the measurement, the Bayes decision will be the closest integer to $\langle n(g) \rangle$, the mean of $p(n|g)$. This is shown in figure 4.3.1.

Now let us compare the quadratic-cost and the minimum-probability-of-error decision rules. Examination of figure 4.2.5 shows that $p(n|g)$ is nearly symmetrical in n when the response is less than about 12. We thus expect these two decision rules to be the same for this range of response. For greater response, $p(n|g)$ begins to show a tail at large values of n, so we expect the minimum-probability-of-error decision to be slightly less than the quadratic-cost decision. In the following table we compare the decisions for these two cost functions; the anticipated deviation appears for a response of 13 counts.

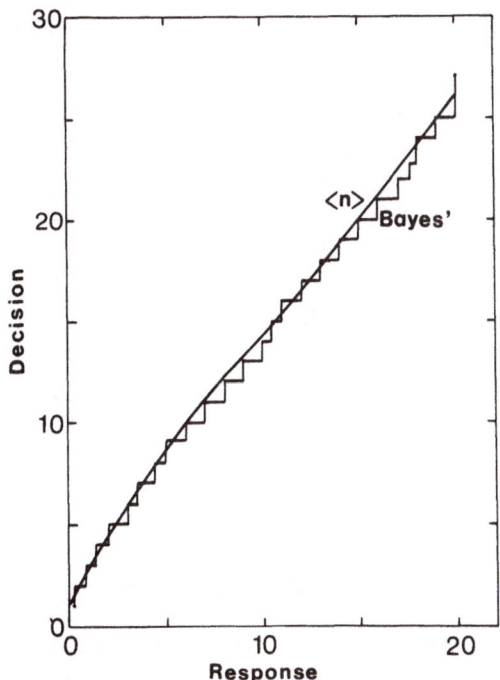

Figure 4.3.1 Bayes decision for a quadratic cost function, compared with the average, $\langle n(g) \rangle$, of $p(n|g)$. Evaluated for a one-detector assay of a plutonium-containing sample.

Response	Bayes Decision	
	Quadratic Decision	Min. Prob. of Error Decision
2	5	5
4	7	7
6	10	10
8	12	12
10	14	14
12	16	16
13	18	17
14	19	18
16	21	21
18	24	23
20	27	26

4.3.7 Estimating Continuous Parameters

In many assay applications the measurements must be used to infer the value of a continuous parameter, such as the total mass of analyte in the sample. In this section we shall briefly indicate how the Bayesian approach may be applied.

It is necessary to know the joint probability density of the response, g, and of the parameter to be estimated, x. This joint density is denoted $p(g, x)$. Furthermore, a cost function $c(x', x)$ must be selected. This function expresses the penalty associated with estimating the parameter as x', when in fact its value is x. We shall consider the squared-error cost function:

$$c(x', x) = (x - x')^2 \qquad . \qquad (4.3.46)$$

The risk is evaluated as the average of the cost over the entire range of possible values of x and of g. That is,

$$R = \int \int c(x', x) p(g, x) \, dx \, dg \qquad . \qquad (4.3.47)$$

Employing the squared-error cost function this may be written as

$$R = \int p(g) \int (x - x')^2 p(x|g) \, dx \, dg \qquad (4.3.48)$$

We wish to choose x' — the estimate of x — so as to minimize the risk. Differentiating R we find

$$\frac{dR}{dx'} = -2 \int p(g) \int x p(x|g) \, dx \, dg$$
$$+ 2 \int p(g) x' \int p(x|g) \, dx \, dg$$
$$= 2 \int p(g)(x' - \langle x|g \rangle) \, dg$$

where $\langle x|g \rangle$ is the conditional mean of x. Since the second derivative of R is positive, we find that R is minimized by choosing the estimate as

$$x'(g) = \langle x|g \rangle = \int x p(x|g) \, dx \qquad . \qquad (4.3.49)$$

This analysis can be readily extended to the case where n parameters $x_1, \ldots, x_n$ are to be estimated on the basis of m measurements $g_1, \ldots, g_m$. The squared-error cost function is

$$c(x'_1, \ldots, x'_n; x_1, \ldots, x_n) = \sum_{i=1}^{n} (x_i - x'_i)^2 \qquad . \qquad (4.3.50)$$

Following a development analogous to our previous discussion we find that the estimate of x_i, given the measurements $g_1, \ldots, g_m$, is

$$x_i'(g_1, \ldots, g_m)$$
$$= \langle x_i | g_1, \ldots, g_m \rangle$$
$$= \int x_i p(x_1, \ldots, x_n | g_1, \ldots, g_m) \, dx_1 \cdots dx_n \quad (4.3.51)$$

We shall have occasion to use these results in Chapter 6.

4.4 NEYMAN-PEARSON DECISION THEORY

The Bayes decision theory which we discussed in the previous section is a powerful and versatile tool. An attractive aspect of the Bayes approach is the flexibility with which the cost function incorporates widely varying decision priorities. A drawback of the Bayes theory, however, is that it requires knowledge of the *a priori* probability, $p(n)$, of occurrence of n particles in the sample. In this section we shall discuss an alternative approach — the Neyman-Pearson theory — which avoids this difficulty and yet retains, to a certain degree, the versatility of the Bayes theory. A general formulation would be unnecessarily complicated. Instead we will proceed directly to two examples which will amply illustrate the general technique.

4.4.1 Threshold Detection

The basic input to the Neyman-Pearson theory is $p(g|n)$, the probability density of the response conditioned by the number of source particles. The aim, as in the Bayes case, is to divide the set G of all possible responses, into disjoint decision-regions G_m: the response g is interpreted as arising from m particles if g belongs to G_m.

We shall assume that the number of source particles may be any value from zero up to N. In the threshold detection problem we wish to divide G into two decision regions G_0 and G_1 which respectively identify whether or not the number of source particles exceeds a given value, M, which is less than N. The construction of these decision regions is to be subjected to a constraint.

Given that not more than M particles are in the sample, the likelihood that the response g will occur is expressed by

$$p_0(g) = \frac{\displaystyle\sum_{m=0}^{M} p(g|m)}{\displaystyle\int_G \sum_{m=0}^{M} p(g|m)\, dg} \qquad (4.4.1)$$

$$= \frac{1}{M+1} \sum_{m=0}^{M} p(g|m) \qquad (4.4.2)$$

Similarly, given more than M source particles in the sample, the likelihood that the response g will occur is

$$p_1(g) = \frac{1}{N-M} \sum_{m=M+1}^{N} p(g|m) \qquad (4.4.3)$$

We note that $p_0(g)$ and $p_1(g)$ are not strictly conditional probability densities; to calculate those densities would require knowledge of the *a priori* distribution of n.

The likelihood of falsely identifying $m \le M$ source particles is

$$P_0 = \int_{G_1} p_0(g)\, dg \qquad (4.4.4)$$

Similarly, the likelihood of mistakenly interpreting the response from more than M particles is

$$P_1 = \int_{G_0} p_1(g)\, dg \qquad (4.4.5)$$

It is worthwhile to note the symmetry of P_0 and P_1. This symmetry will enable easy formulation of the decision rule.

We wish to choose the decision regions G_0 and G_1 so that P_1 is a minimum, and so as to satisfy the following constraint on P_0:

$$P_0 = c \qquad (4.4.6)$$

where c is a non-negative quantity less than unity which we are free to choose.

The method of Lagrange multipliers provides a direct solution of this problem. Let us define

$$J = P_1 + \lambda(P_0 - c) \qquad (4.4.7)$$

Clearly, minimizing J while satisfying eq.(6) achieves a minimization of P_1.

Using eqs.(4) and (5) we may express J as

$$J = \int_{G_0} p_1(g)\, dg + \lambda \int_{G_1} p_0(g)\, dg - \lambda c \qquad (4.4.8)$$

Since G_0 and G_1 are disjoint, they are related by

$$G_1 = G - G_0 \qquad (4.4.9)$$

Hence eq.(8) can be written

$$J = \int_{G_0} \big(p_1(g) - \lambda p_0(g)\big)\, dg + \lambda(1 - c) \qquad (4.4.10)$$

We shall presently show that λ is non-negative. Using this fact, it is evident that J is minimized by the decision rule:

$$g \in G_0 \qquad (4.4.11)$$

if and only if

$$\frac{p_1(g)}{p_0(g)} < \lambda \qquad (4.4.12)$$

The value of λ determines the division of G into the decision regions G_0 and G_1. Thus λ must be chosen so as to satisfy the constraint, eq.(6). Since g is a random variable, the likelihood ratio

$$L(g) = \frac{p_1(g)}{p_0(g)} \qquad (4.4.13)$$

is also a random variable. Let $p_0(L)$ denote the probability density of L, given that the number of source particles does not exceed M. Then the constraint may be expressed:

$$P_0 = \int_{\lambda}^{\infty} p_0(L)\,dL = c \qquad (4.4.14)$$

since, when $L(g) \geq \lambda$, the response g is assigned to G_1. Eq.(14) may be solved for λ in terms of the constraint c. Furthermore, we see that λ must be non-negative because $c < 1$ and since $p_0(L)$ is zero for $L < 0$.

4.4.2 Maximum Likelihood Estimation

The decision rule which we developed for threshold detection can be readily generalized to problems involving multiple hypotheses or estimation of a continuous parameter. Consistent with the Neyman-Pearson approach, we shall rely only on knowledge of the conditional probability density of the response given the quantity of analyte, $p(g|x)$. The variable x may be discrete, as in the case of estimating the number of identical source particles, or it may be continuous, as in estimating the mass of analyte.

The threshold decision rule hinges on comparing the conditional density of the response for the two alternative hypotheses. Referring to eq.(12) we see that when $p_1(g)$ is sufficiently greater than $p_0(g)$, we interpret the measurement as arising from condition 1; conversely, when $p_0(g)$ is sufficiently greater than $p_1(g)$ we make the opposite decision. When we must select one from among a range (or continuum) or alternatives, this comparative procedure has a natural extension: we interpret the response g as arising from the most likely alternative. That is, g is interpreted as arising from the value of x which maximizes $p(g|x)$ on the set of values of x. This "maximum likelihood estimate" is denoted in this section as $x'(g)$. In general, the maximum likelihood estimate may be defined by

$$p\big(g|x'(g)\big) = \underset{x}{\text{maximum}}\, p(g|x) \qquad . \qquad (4.4.15)$$

If x is a continuous variable and if $p(g|x)$ is smooth and uni-modal, then $x'(g)$ may be defined as

$$\left| \frac{\partial p(g|x)}{\partial x} \right|_{x'(g)} = 0 \qquad . \qquad (4.4.16)$$

4.5 DIRECT PROBABILISTIC CALIBRATION

In the previous two sections we have developed two versatile approaches for generating decision-rules for interpreting the measured response of a sample. These decision theories — of Bayes and of Neyman and Pearson — enable incorporation of a wide range of considerations in the definition of the decision rule. In particular, varying penalties for incorrect interpretation can be adopted in accordance with practical or physical considerations of the particular application in question. We have seen that, in the Bayes theory, the quadratic cost function generates a decision rule which is essentially the mean, $\langle n(g) \rangle$, of the probability density of the number of source particles conditioned by the response. That is, the quadratic cost function leads one to interpret the response g as the mean of $p(n|g)$. This function $\langle n(g) \rangle$ may be called a "probabilistic calibration curve". This decision rule is widely employed because the average of $p(n|g)$ has a clear physical meaning and because the quadratic cost function seems a reasonable penalty for error. A further advantage of the probabilistic calibration $\langle n(g) \rangle$ is that the uncertainty of the interpretation which is due to the random spatial distribution of the source material is succinctly expressed by the standard deviation of the probability density $p(n|g)$. We have also shown how the statistical uncertainty of the measurement may be incorporated in evaluation of the uncertainty of the decision.

Alternatively, assigning zero cost to a correct decision and unit cost to any incorrect decision generates a decision rule which minimizes the average probability of error. Other cost functions yield different decision rules with various particular features. In each case, the Bayes theory requires knowledge of the probability density of the measurement conditioned by the number of source particles, $p(g|n)$, and knowledge of the *a priori* probability of the number of source particles, $p(n)$. In the Neyman-Pearson approach somewhat similar decision-rules can be generated, without requiring knowledge of $p(n)$. However, we have seen in section 4.1 that it is not a trivial matter to compute the conditional density $p(g|n)$, and the computational burden of generating the decision regions G_m is also not negligible.

In light of these considerations, we shall devote this section to the presentation of a technique for evaluating a probabilistic calibration curve directly from the basic formulation of the problem. That is: without the necessity of evaluating the probability density $p(g|n)$. In our general formulation of this technique we shall assume that the response function satisfies the assumptions of linearity in time and mass defined in section 2.2. Furthermore we shall assume that source deposits in disjoint regions of the sample are independently positioned. In our general formulation we need not assume that the source deposits are identical, nor that the spatial distribution is uniformly random over the sample. Following the general formulation we shall discuss an application to the problem of prospecting for uranium.

4.5.1 General Formulation — One Detector

The basic component in our direct construction of a probabilistic calibration is the response function $f(x, y)$ which represents the response produced in unit time in a detector located at point y by a unit of source material located at position x. Let $r(x)\,dv$ represent the mass of source material in the differential volume dv at point x in the sample. Thus $r(x)\,dv$ is a random variable, and is an implicit function of the average concentration of source material in the sample and of the size-distribution of source deposits. At this point we are placing no restrictions on the variability of $r(x)\,dv$, though we shall assume that various moments and cross-moments of $r(x)\,dv$ can be evaluated. These expectations will also be functions of the average concentration of the source material and of the size-distribution of the source deposits.

For a particular spatial distribution of source material, $r(x)$, the response of a single detector will be

$$g_1 = t \int f(x,y)r(x)\,dv \qquad (4.5.1)$$

where the integral is over the entire volume of the sample and t is the duration of the measurement. Furthermore, since we assume that source deposits in disjoint regions of the sample are statistically independent, the average response is simply

$$\langle g_1 \rangle = t \int f(x,y)\langle r(x)\,dv \rangle \qquad (4.5.2)$$

Similarly, the spatial variance of the response due to random spatial distribution of the source material may be expressed

$$\mathrm{var}_{sp}(g_1) = \langle (g_1 - \langle g_1 \rangle)^2 \rangle$$
$$= t^2 \int \int f(x,y)f(x',y)\mathrm{cov}\big(r(x)\,dv\,,\,r(x')dv'\big)$$
$$(4.5.3)$$

where $\mathrm{cov}\big(r(x)\,dv, r(x')dv'\big)$ represents the covariance of the quantity of source material in the volumes dv and dv', and is defined as

$$\mathrm{cov}\big(r(x)\,dv\,,\,r(x')dv'\big)$$
$$= \langle r(x)\,dv\,r(x')dv' \rangle - \langle r(x)\,dv \rangle\langle r(x')dv' \rangle \qquad (4.5.4)$$

Since we are assuming that disjoint volumes of source material are statistically independent, the covariance of the mass in dv and dv' vanishes unless these differential volumes coincide. Thus eq.(3) simplifies to

$$\text{var}_{sp}(g_1) = t^2 \int f^2(x,y)\text{var}\big(r(x)\,dv\big) \qquad (4.5.5)$$

where $\text{var}(r(x)\,dv)$ is the variance of the source mass in the volume dv.

Now $\text{var}_{sp}(g_1)$ is the variance of the response of one detector, due to the random spatial distribution of the source material. It is often necessary to include the effect of the statistical uncertainty of the response. For Poisson counting statistics, the variance of the number of counts equals the mean. Hence the statistical uncertainty of the response of a single detector is

$$\text{var}_{st}(g_1) = \langle g_1 \rangle \qquad (4.5.6)$$

We may usually assume that the processes generating the statistical uncertainty and the spatial uncertainty are independent. Hence the overall variance of the response is

$$\text{var}(g_1) = \text{var}_{sp}(g_1) + \text{var}_{st}(g_1) \qquad (4.5.7)$$

4.5.2 General Formulation — Multiple Detectors

Let us now consider k detectors located at points $y_1, \ldots, y_k$. The response function of the i-th detector is $f(x, y_i)$. For a given spatial distribution, the sum of the k responses will be

$$g_k = t \sum_i \int f(x, y_i) r(x)\,dv \qquad (4.5.8)$$

The response, averaged over the possible spatial distributions, is

$$\langle g_k \rangle = t \sum_i \int f(x, y_i) \langle r(x)\,dv \rangle \qquad (4.5.9)$$

and the spatial variance of the response is

$$\mathrm{var}_{sp}(g_k)$$

$$= t^2 \sum_i \sum_j \int f(x, y_i) f(x, y_j) \mathrm{var}\big(r(x) \, dv\big) \quad (4.5.10)$$

As before, the statistical variance of g_k is

$$\mathrm{var}_{st}(g_k) = \langle g_k \rangle \qquad\qquad (4.5.11)$$

and the overall variance of the sum of the k measurements is

$$\mathrm{var}(g_k) = \mathrm{var}_{sp}(g_k) + \mathrm{var}_{st}(g_k) \qquad\qquad (4.5.12)$$

4.5.3 Example: U Prospecting — Assay of a Thick Deposit

Uranium deposits of potential economic importance contain uranium concentrations ranging from several hundred to several thousand ppm U. These deposits occur in a wide variety of geological circumstances, and display a range of different shapes and sizes. The search for uranium deposits typically employs geological analysis [10], analysis of soil and ground-water samples [11], airborne gamma-ray survey [12], and bore-hole logging with passive-gamma [13] or neutron-activation techniques [14]. The most widespread bore-hole logging technique employs passive gamma assay with a single sodium iodide detector.

In many uraniferous deposits the uranium occurs in homogeneous layers whose thickness greatly exceeds the mean free path of the detected radiation. In such a circumstance the detector may be accurately calibrated with the aid of a standardized laboratory model of the uraniferous layer [15]. In other geological environments of economic importance the uraniferous deposits are far from homogeneous, and occur randomly as nodules or pods in irregular veins [16]. In this case the calibration of a bore-hole logging apparatus is confronted with a difficult problem of spatial randomness of the source material.

In the present example we shall illustrate the application of the direct probabilistic calibration to the problem of the assay of an heterogeneous uranium vein. We shall consider a thick layer of small uniform nodules of unknown size. We shall assume that the uranium nodules are randomly and independently distributed around the fixed sodium iodide detector. We shall ignore the finite size of the bore-hole in which the detector is located, other than to assume that a uranium nodule may not be any closer to the detector than some minimum distance x_0.

By assuming the uranium nodules to be randomly and independently located, the probability of n nodules occurring in a volume V is given by the Poisson distribution as

$$p(n, V) = \frac{e^{-NV}(NV)^n}{n!} \qquad (4.5.13)$$

where N is the average number density of uranium nodules. We immediately obtain the average and variance of the number $n(V)$ of nodules in the volume V as

$$\langle n(V) \rangle = \text{var}\big(n(V)\big) = NV \qquad (4.5.14)$$

The response function of the NaI detector for measurement of the 1.765 MeV gamma radiation produced in the decay chain of ^{238}U is

$$f(x) = \frac{C}{4\pi x^2} e^{-\mu x} \qquad \text{counts/g sec} \qquad (4.5.15)$$

where μ is the linear absorption coefficient of the matrix, C is a product of various constants and x is the distance from the uranium nodule to the detector.

We are now able to evaluate the average response of a single detector, eq.(2). Employing spherical coordinates, the average number of nodules in a spherical shell of thickness dx is

$$\langle r(x)\, dv \rangle = 4N\pi x^2\, dx \qquad (4.5.16)$$

Thus the average response becomes

$$\langle g_1 \rangle = \frac{tCMN}{\mu} e^{-\mu x_0} \qquad (4.5.17)$$

where M is the mass of a uranium nodule. Since the uranium nodules have uniform random spatial distribution, the average response precisely equals the response that would be obtained from a homogeneous distribution of uranium at the same concentration.

In a similar fashion we may obtain the spatial variance of the response, from eqs.(5), (14) and (15) as

$$\mathrm{var}_{sp}(g_1) = \frac{(tCM)^2 N}{4\pi} \int_{x_0}^{\infty} \frac{1}{x^2} e^{-2\mu x}\, dx \qquad (4.5.18)$$

This can be represented in a somewhat more convenient fashion by employing the exponential integral for real arguments:

$$E_n(x) = \int_{1}^{\infty} \frac{e^{-xt}}{t^n}\, dt \quad , \quad n = 0, 1, 2, \ldots \quad ; \quad x > 0 \qquad (4.5.19)$$

By a change of variables one obtains

$$x^{1-n} E_n(cx) = \int_{x}^{\infty} \frac{e^{-cy}}{y^n}\, dy \qquad (4.5.20)$$

Thus the spatial variance becomes

$$\mathrm{var}_{sp}(g_1) = \frac{(tCM)^2 N}{4\pi x_0} E_2(2\mu x_0) \qquad (4.5.21)$$

The uranium concentration, w, is related to the nodule size, M, and to the average number density, N, by

$$w = MN \qquad g/cm^3 \qquad (4.5.22)$$

Employing this relation the average and spatial variance of the response become

$$\langle g_1 \rangle = \frac{tCw}{\mu} e^{-\mu x_0} \qquad (4.5.23)$$

$$\mathrm{var}_{sp}(g_1) = \frac{(tCw)^2}{4\pi x_0 N} E_2(2\mu x_0) \qquad (4.5.24)$$

We see that the average uranium concentration, w, uniquely determines the average response, $\langle g_1 \rangle$. In other words, eq.(23) constitutes a probabilistic calibration of the measurement. On the other hand, the spatial variance depends not only on the average concentration, but also on the number density of the uranium nodules. As the number density decreases (at constant average concentration) the probability of unusual spatial distributions increases, causing the spatial variance to grow.

From eqs.(11) and (23) we can express the statistical variance of the response as

$$\mathrm{var}_{st}(g_1) = \frac{tCw}{\mu} e^{-\mu x_0} \qquad (4.5.25)$$

We see that the statistical variance depends on the average uranium concentration, and on the duration of the measurement.

In uranium assay, as in many other applications, it is often important to include consideration of the statistical uncertainty introduced by background radiation. A source of background radiation of particular importance in uranium assay is ^{232}Th and its decay daughters. In particular, ^{208}Tl produces gamma radiation at 2.61 MeV, whose scatter interferes with the 1.765 MeV radiation from the uranium decay chain. The statistical variance of g_1 due to the background radiation is related to the average intensity of the background radiation as

$$\mathrm{var}_{bg}(g_1) = \langle g_{bg} \rangle \qquad (4.5.26)$$

Let the ratio of background radiation to uranium radiation be

$$b = \frac{\langle g_{bg} \rangle}{\langle g_1 \rangle} \qquad (4.5.27)$$

Thus the total variance of g_1 is

$$\text{var}(g_1) = \text{var}_{sp}(g_1) + (1 + b)\text{var}_{st}(g_1) \qquad (4.5.28)$$

Let $\sigma(g_1)$ be the total standard deviation of g_1. The relative error of the response, $\sigma(g_1)/\langle g_1 \rangle$, is of particular interest for assessing the affect of spatial and statistical uncertainty. This may be expressed as

$$\frac{\sigma(g_1)}{\langle g_1 \rangle}$$

$$= \left(\frac{\mu^2 M}{4\pi x_0 w} E_2(2\mu x_0) e^{2\mu x_0} + \frac{(1+b)\mu}{tCw} e^{\mu x_0} \right)^{1/2}$$

The relative error of the response depends on the average uranium concentration, as well as on the nodule size (or number density), the duration of measurement, and the ratio of background to uranium radiation. In figure 4.5.1 we plot the relative error versus the uranium weight fraction (mass concentration divided by the matrix density) for various nodule sizes. The measurement duration is 10 seconds and the background-to-peak ratio, b, equals unity. The dashed line is the relative error due to statistical uncertainty (including background), while the solid lines represent the total relative error. We see that for very small nodules, the dominant source of error is statistical, while for large nodules the spatial uncertainty is of overriding importance. Furthermore for sufficiently large nodules the relative error is quite large even at very high average concentration. This indicates that, for sparsely dispersed nodules, the probabilistic calibration is likely to be quite unreliable. We shall raise this point again when we return, in Chapter 5, to the problem of assay-system design on the basis of probabilistic performance criteria.

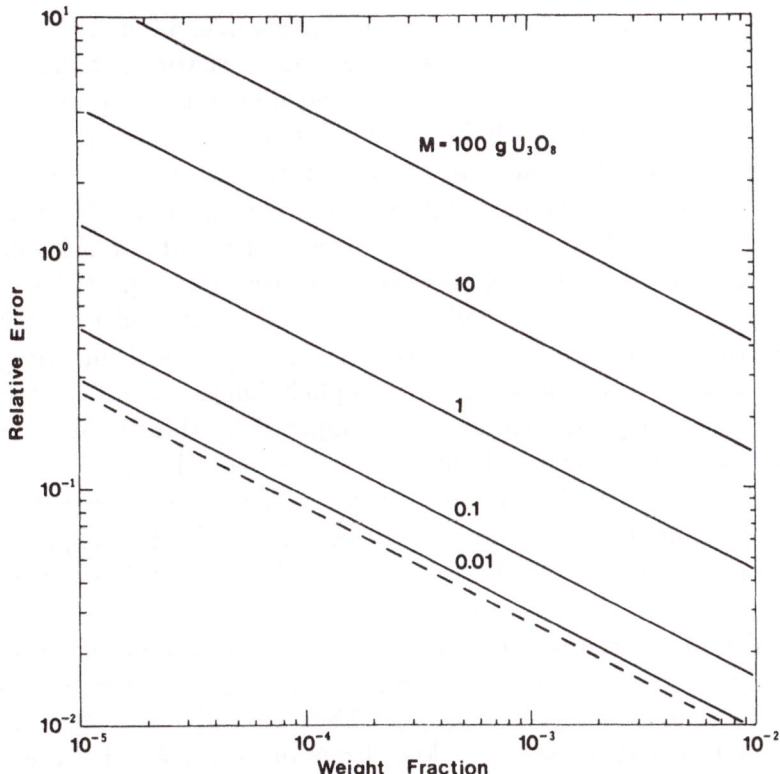

Figure 4.5.1 The total relative error (solid lines) and the statistical relative error (dashed line) versus the uranium weight fraction.

4.6 SUMMARY

The main theme of this Chapter is the development of various techniques for the interpretation of the measurement of spatially random materials. Three different classes of solutions to this problem are discussed: Bayes decision theory, Neyman-Pearson decision theory and direct probabilistic calibration. Each approach is based on assessment, in one way or another, of the relative probability or likelihood of different spatial distributions.

Our first task was to develop an expression for the conditional probability density of the measurement vector, g, given the number

n_i of source particles of mass m_i in the sample: $p(g|n_1, \ldots, n_w)$. While we restricted our discussion to random spatial distributions of discrete source particles, we imposed no explicit limitation on the range of masses which the source particles may assume, nor did we require that the source particles be identical. Thus continuous mass distributions can be approximated with arbitrary accuracy. Indeed we found in the example in section 4.1.5 that $p(g|n)$ is quite sensitive to the degree of subdivision of the source particles. In order to construct this conditional probability we had to require that the response functions satisfy the assumptions of linearity in time and linearity in mass. (An implicit limitation on the variability of source-particle masses is imposed by this assumption). Furthermore we assumed that the source particles are independently located in the sample. However, it was not necessary to assume that the distribution of source particles in the sample was uniformly random. Rather, we showed that non-uniform spatial distribution of the source particles is readily handled.

Our next step was the construction of the conditional probability density of the total source mass, x, given the value g of the measurement vector: $p(x|g)$. The construction of this conditional density is based on knowledge of $p(g|n)$ and knowledge of the marginal or *a priori* distribution of the number of source particles $p(n)$, where n is a vector in the case of non-identical source particles. We discussed the applicability of the Poisson distribution for representing the marginal distribution $p(n)$ in certain circumstances. Then we showed that the likelihood function may be used as a plausible substitute for $p(x|g)$ when knowledge of $p(n)$ is lacking. Finally we showed how statistical uncertainty can be incorporated in the probability distribution $p(x|g)$.

In our subsequent discussion of the Bayes, Neyman-Pearson and direct probabilistic calibration techniques we restricted ourselves to consideration of identical source particles. Generalization to source particles of variable mass does not present any fundamental difficulty.

The Bayes decision theory is based on assigning a cost C_{mn} to deciding that any given measurement arose from m source particles when in fact n particles are in the sample. The set of all possible values of the measurement vector is divided into decision regions, $G_1, G_2, \ldots$, so that a measurement g is interpreted as arising from m source particles if g belongs to G_m. The overall average cost or risk, R, is expressed as a function of the decision regions, which are then chosen so as to minimize R. The implementation of the Bayes approach requires knowledge of both $p(g|n)$ and $p(n)$. Various different cost functions C_{mn} are examined, and it is shown that sometimes quite different decision rules are generated.

The Neyman-Pearson decision theory is developed next. It is similar to Bayes' approach in that decision regions are generated which minimize or maximize the probabilities or likelihoods of various decision errors or successes. Instead of specifying a cost function for penalizing incorrect decision, the Neyman-Pearson approach allows one to choose the decision regions in accordance with constraints on the probabilities or likelihoods of various decision errors or successes. Implementation requires knowledge of $p(g|n)$ but not of $p(n)$.

The final decision scheme developed is the technique of direct probabilistic calibration. We discuss a procedure for deriving the mean and standard deviation of the measurement vector directly from the formulation of the assay problem. One must know the response functions of the detectors and the first two moments of the spatial distribution of the source material. No knowledge of the probability densities $p(g|n)$, $p(n)$ or $p(n|g)$ is required. This approach is considerably less versatile than the others, however its implementation requires less complete stochastic understanding of the assay problem and usually calls for less computational effort.

From the various examples discussed in this Chapter we have strengthened the feeling that the deterministic measure of performance developed in Chapters 2 and 3 yields a conservative estimate of the mass resolution capability of the assay system. The utility of the deterministic performance criterion is in its ease of application to a large number of proposed assay-system designs,

and its succinct assessment of their merits. It is useful as a design tool for narrowing the range of assay-systems which need
be subjected to more thorough scrutiny. Our development, in
the present Chapter, of tools for probabilistic interpretation of
the measurement of spatially random materials, lays the ground-
work for our consideration, in the next Chapter, of probabilistic
measures of performance. These probabilistic criteria provide the
means for detailed comparison of selected proposed assay-system
designs, leading to a final design decision.

NOTES

[1] 1. F. A. Fry, B. M. R. Green, A. Knight and D. R. White, A
Realistic Chest Phantom for the Assessment of Low Energy Emit-
ters in Human Lungs, 4-th Intl. Conf. of the Intl. Rad. Protection
Soc., Vol. 2, pp 475-8, Paris, 1977.
2. W. W. Parkinson, jr., R. E. Goans and W. M. Good, Realistic
Calibration of Whole-Body Counters for Measuring Plutonium,
Intl. Atomic Energy Agency, Conf. on Natl. and Intl. Standard-
ization of Rad. Dosimetry, Vol. III, pp 155-66, Vienna, 1977.

[2] For extensive discussion of reaction rates and penetration distributions in pores see
1. A. Wheeler, Reaction Rates and Selectivity in Catalyst Pores,
in *Catalysis*, Vol. II, ed. by P. H. Emmett, Reinhold, 1955.
2. A. Wheeler and A. J. Robell, Performance of Fixed-Bed Cat-
alytic Reactors with Poisson in the Feed, *J. of Catalysis*, 13: 299-
305(1969).

[3] This assumption is physically reasonable since the response
function is expected to be continuous and smooth. However, if
the probability density does not exist the only alteration needed
in our subsequent derivation is that the integrals involving the
probability density must be changed to Lebesque integrals with
the probability distribution P as the measure.

[4] See ref. [4] of Chapter 2.

[5] See ref. [15] no. 1 of Chapter 2.

[6] J. L. Melsa and D. L. Cohn, *Decision and Estimation Theory*, McGraw-Hill, 1978.

[7] For a fundamental discussion of likelihood ratios see J. L. Doob, *Stochastic Processes*, John Wiley, 1953. For a briefer and simpler treatment see ref. [4] of Chapter 2.

[8] If $p(n)$ is a non-increasing sequence we can prove an inequality on the ordinary moments of $p(n|g)$ and $L(n|g)$. For simplicity let us assume that the source particles are all of the same mass, so that n is simply a number rather than a vector. Let us denote

$$P_n = p(n|g) \qquad \text{and} \qquad L_n = L(n|g)$$

and

$$a = \frac{\sum\limits_n p(g|n)}{\sum\limits_n p(g|n)p(n)} \quad .$$

Then

$$P_n = ap(n)L_n \quad .$$

The ordinary moments of the distributions L_n and P_n are

$$u(r) = \sum n^r L_n$$
$$v(r) = \sum n^r P_n = \sum ap(n)n^r L_n$$

Now, Chebyshev's inequality states that, if $q_0, q_1, q_2, \ldots$ is a probability distribution and if $x_0, x_1, x_2, \ldots$ and $y_0, y_1, y_2, \ldots$ are non-negative oppositely ordered sequences, then

$$\sum q_i x_i y_i \leq \left(\sum q_i x_i \right) \left(\sum q_i y_i \right)$$

For a complete statement of Chebyshev's inequality see: G. H. Hardy, J. E. Littlewood and G. Polya, *Inequalities*, Cambridge University Press, 1951.

For $r > 0$ the sequence $0^r, 1^r, \ldots, n^r, \ldots$ is increasing. If $p(n)$

is non-increasing then the sequences $ap(n)$ and n^r are oppositely ordered. From Chebyshev's inequality we see that

$$\sum ap(n)n^r L_n \leq \left(\sum n^r L_n\right)\left(\sum ap(n)L_n\right)$$

But

$$\sum ap(n)L_n = \sum P_n = 1 \qquad .$$

Hence

$$v(r) \leq u(r)$$

which proves that, if $p(n)$ is non-increasing, the ordinary (not central) moments of P_n do not exceed the ordinary moments of L_n. An important example of a non-increasing sequence $p(n)$ is the Poisson distribution:

$$p(n) = \frac{e^{-x} x^n}{n!} \qquad , \qquad n = 0, 1, 2, \ldots \qquad .$$

When $x \leq 1$, we see that $p(n+1) \leq p(n)$ for all n.

[9] 1. H. L. Van Trees, *Detection, Estimation and Modulation Theory*, Part I, John Wiley, 1968.
2. See reference [6].

[10] International Atomic Energy Agency, Technical Reports Series No. 186, *Gamma-Ray Surveys in Uranium Exploration*, Vienna, 1979.

[11] 1. T. W. Parker, Determination of the Concentration of Uranium in Soil and Stream Sediment Samples Using a High Resolution Energy-Dispersive X-Ray Fluorescence Analyzer, *Int. J. Appl. Rad. Isot.*, 34:273-81 (1983).
2. J. K. Osmond, J. B. Cowart and M. Ivanovich, Uranium Isotope Disequilibrium in Ground Water as an Indicator of Anomalies, *ibid*, pp 283-308.

[12] 1. Q. Bristow, Airborne Gamma-Ray Spectrometry in Uranium Exploration. Principles and Current Practice, *Int. J. Appl.*

Rad. Isot., 34: 199-229 (1983).
For a discussion of the effect of non-uniformity in the distribution of the radioactive material see
2. M. R. Wormald and C. G. Clayton, Observations on the Accuracy of Gamma Spectrometry in Uranium Prospecting, Intl. Atomic Energy Agency Conf. on Exploration for Uranium Ore Deposits, Vienna, 1976, pp 147 - 71.

[13] P. G. Killeen, Borehole Logging for Uranium by Measurement of Natural Gamma-Radiation, *Int. J. Appl. Rad. Isot.*, 34: 231-60 (1983).

[14] 1. D. R. Humphreys *et al*, Uranium Logging with Prompt Fission Neutrons, *Int. J. Appl. Rad. Isot.*, 34:261-8 (1983).
2. L. A. Shope *et al*, The Operation and Life of the Zetatron Neutron Tube in a Borehole Logging Application, *ibid*, pp 269-72.
3. M. R. Wormald and C. G. Clayton, Some Factors Affecting Accuracy in the Direct Determination of Uranium by Delayed Neutron Borehole Logging, Intl. Atomic Energy Agency Conf. on Exploration for Uranium Ore Deposits, Vienna, 1976, pp 427 - 70.

[15] International Atomic Energy Agency, Technical Report Series No. 174, *Radiometric Reporting Methods and Calibration in Uranium Exploration*, Vienna, 1976.

[16] Various aspects of vein deposits are discussed in:
1. Intl. Atomic Energy Agency Conf. on Vein-Type and Similar Uranium Deposits in Rocks Younger Than the Proterozoic, Lisbon, 1979.
In particular, some data on the micro-structure of such deposits is contained in the paper:
2. J. Rimsaite, Chemical and Isotopic Evolution of Radioactive Minerals in Remobilized Vein-Type Uranium Deposits, Saskatchewan, Canada, *ibid*, pp 35 - 46.

CHAPTER 5

PROBABILISTIC DESIGN

5.1 MOTIVATION

In the previous three Chapters we have studied assay-system design and measurement-interpretation as separate tasks in the assay of spatially random materials. In Chapters 2 and 3 we developed and generalized the concept of relative resolution as a measure of performance. The relative resolution is a deterministic design tool in the sense that it accounts for all allowed spatial distributions of the analyte, without consideration of the relative probability of different distributions. The relative resolution is a concise physically meaningful quantity, and it can be formulated algorithmically for convenient computation. In Chapter 4 we turned our attention from assay-system design to probabilistic interpretation of measurement. The basic tool in probabilistic analysis is the conditional probability density. We studied the construction of conditional densities, and we found that extensive stochastic information about the assay problem is required, as well as considerable computational effort.

Chapters 2 to 4 have treated design and interpretation as separate tasks. In this and the next Chapter these tasks are integrated. The aims and means of such an integration are various. In the present Chapter we shall apply probabilistic tools from Chapter 4 to the design analysis. The foremost motivation for a probabilistic design analysis is that the deterministic measure of performance tends to be conservative: to underestimate the resolution capability. The deterministic analysis, by virtue of its computational efficiency, may be used to examine a broad range of design concepts and to narrow the field of design options. The probabilistic design analysis, which requires more detailed formulation of the assay problem and which involves greater computational effort, is employed for final selection of the assay system design. The division of labor between deterministic and probabilistic approaches to design should not be viewed as irrevocable. In some circumstances the deterministic analysis will suffice for final design decisions, while in other cases only a probabilistic analysis is meaningful.

In the present Chapter we shall discuss three quite different approaches to probabilistic design analysis. The techniques differ in generality of application, computational difficulty, and the extent of probabilistic information required about the assay problem.

5.2 RELATIVE ERROR CRITERION

5.2.1 General Considerations

The simplest probabilistic design analysis is based on the concept of relative error: the ratio of the standard deviation to the mean of some quantity of interest. We employed this concept in our discussion of direct probabilistic calibration in section 4.5. We showed that the relative error of the measurement vector, $\sigma(g)/\langle g \rangle$, can be evaluated directly from the definition of the assay problem, without requiring construction of a probability density function. Alternatively, if the probability density of the source mass x conditioned by the measurement vector g is available, one can evaluate the relative error of the source mass, $\sigma(x)/\langle x \rangle$. This gives a somewhat more meaningful performance criterion than the relative error of the measurement vector. In either case the relative errors for alternative designs provide direct comparison of the designs.

5.2.2 Example: U Prospecting — Assay of A Thin Deposit

We shall illustrate the use of the relative error as a tool for designing the assay of a spatially heterogeneous uranium deposit. The uranium occurs as uniform nodules of mass M randomly distributed in a layer of thickness T. A cross section of the layer is

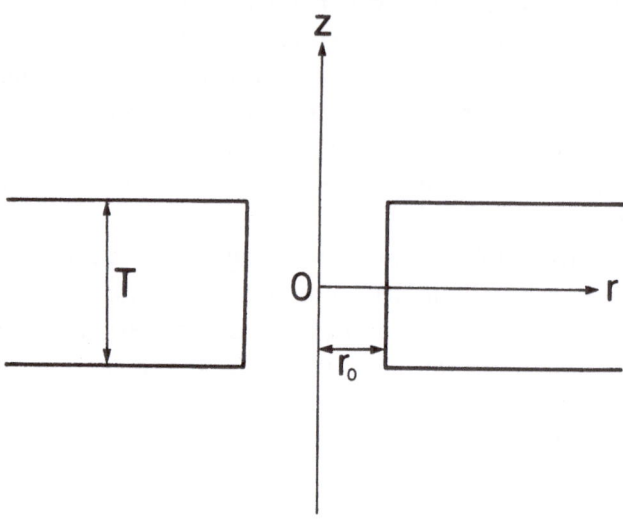

Figure 5.2.1 Cross section of the uraniferous layer of thickness T with bore-hole radius r_0.

shown in figure 5.2.1. The upper and lower surfaces of the uraniferous layer are parallel, and the horizontal extent of the layer is much greater than the mean free path of the 1.765 MeV gamma radiation which is to be measured. The bore hole penetrates the deposit at 90 degrees to its surface. The radius of the bore is r_0.

The 1.765 MeV gamma radiation is detected passively with a NaI detector. Measurements are to be performed at J locations along the axis of the bore (z-axis in figure 5.2.1). The j-th measurement is located at point z_j, and its response in unit time to a point source whose cylindrical coordinates are (r, θ, z) is

$$f(r, \theta, z, z_j) = \frac{C}{4\pi R^2(r, \theta, z, z_j)} e^{-\mu R(r, \theta, z, z_j)} \qquad (5.2.1)$$

where $R(r, \theta, z, z_j)$ is the distance from point (r, θ, z) to the position of the j-th measurement, and μ is the linear absorption coefficient of the matrix. We have made the simplifying assumption that the absorption coefficient of material in the bore hole equals that of the uraniferous matrix.

It has been decided not to exploit the full information available from the J measurements. Rather, in the interest of simplicity of data analysis, only the average of the J measurements will be retained. This average is denoted $g(J)$. Furthermore, a total duration of D seconds is allotted for all of the J measurements: each measurement is performed during D/J seconds. The aim of our design analysis is to choose the number of measurements which should be performed, and to decide on the best location of these measurements. Thus we must evaluate the relative error of $g(J)$ for various values of J and of the measurement positions.

The number density of uranium nodules at point (r, θ, z) is $n(r, \theta, z)$. Assuming that the uranium nodules are all of uniform mass M, we may express the average of J measurements located along the bore hole by

$$g(J) = \frac{1}{J} \sum_{1}^{J} \int_{-T/2}^{T/2} \int_{r_0}^{\infty} \int_{0}^{2\pi} \frac{CDM}{4\pi J} \frac{e^{-\mu R(r,\theta,z,z_j)}}{R^2(r,\theta,z,z_j)} n(r,\theta,z) \, d\theta \, r dr \, dz$$

(5.2.2)

The number of nodules in any given volume is a random variable with a Poisson distribution. Consequently the mean number of nodules in a differential volume at point (r, θ, z) is

$$\langle n(r,\theta,z) \, d\theta \, r dr \, dz \rangle = N \, d\theta \, r dr \, dz \qquad (5.2.3)$$

where N is the average number density of uranium nodules. Furthermore, the covariance of the number of nodules in two different (non-overlapping) differential volumes is zero. The variance of the number of nodules in a differential volume is equal to the mean:

$$\text{var}\big(n(r,\theta,z) \, d\theta \, r dr \, dz\big) = N \, d\theta \, r dr \, dz \quad . \qquad (5.2.4)$$

In light of these relations we find the spatial average of the mean, $g(J)$, of J measurements to be

$$\langle g(J) \rangle = \frac{CDw}{4\pi J^2} \sum_1^J \int_{-T/2}^{T/2} \int_{r_0}^{\infty} \int_0^{2\pi} \frac{e^{-\mu R(r,\theta,z,z_j)}}{R^2(r,\theta,z,z_j)} \, d\theta \, r dr \, dz \qquad (5.2.5)$$

where $w = MN$ is the average uranium concentration. Also, the spatial variance of the average, $g(J)$, of J measurements is

$$var_{sp}\big(g(J)\big)$$

$$= \left(\frac{CD}{4\pi J^2} \right)^2 Mw \int_{-T/2}^{T/2} \int_{r_0}^{\infty} \int_0^{2\pi} \sum_{j=1}^{J} \sum_{k=1}^{J}$$

$$\times \frac{\exp\big(-\mu R(r,\theta,z,z_j) - \mu R(r,\theta,z,z_k)\big)}{R^2(r,\theta,z,z_j) \, R^2(r,\theta,z,z_k)} \, d\theta \, r dr \, dz \,(5.2.6)$$

The statistical uncertainty of the average measurement $g(J)$ equals the mean:

$$var_{st}\big(g(J)\big) = \langle g(J) \rangle \quad . \qquad (5.2.7)$$

In addition to the statistical uncertainty of the measurement there is uncertainty in determination of the background. Let b be the ratio of background to uranium count rates. The total variance of $g(J)$ may be expressed as the sum of three terms — spatial, statistical and background — as in eq.(4.5.28):

$$var\big(g(J)\big) = var_{sp}\big(g(J)\big) + (1+b)var_{st}\big(g(J)\big) \qquad (5.2.8)$$

We shall now numerically compare one- and two-measurement designs. The total duration of measurement D is 20 seconds, the uranium nodule mass M is 1 gram, the average uranium concentration w is 0.1 g/cm^3, the radius r_0 of the bore hole is 3 cm, and

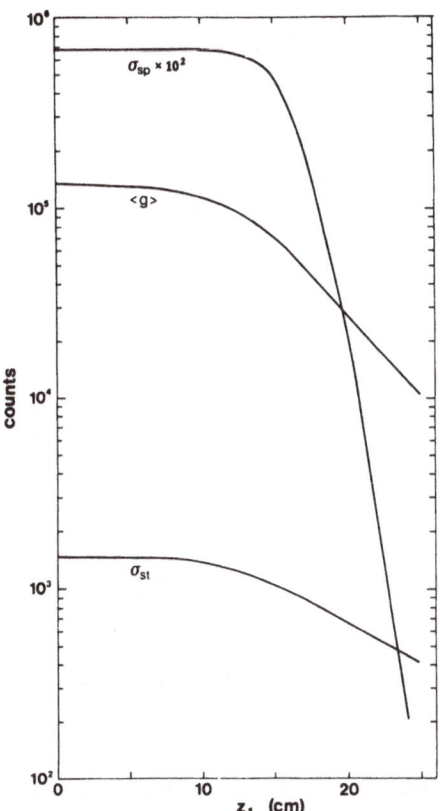

Figure 5.2.2 Average measurement and statistical and spatial standard deviations versus the axial position of a single measurement. Note that the spatial standard deviation has been multiplied by 100.

the ratio of background to uranium count rate b is 3. The thickness of the uraniferous layer, T, is 30 cm. The constant C equals 1.6×10^4, and the linear absorption coefficient is 0.13 1/cm.

In figure 5.2.2 we show the average measurement and the statistical and the spatial standard deviations versus the axial position of the measurement. Note that the spatial standard deviation has been multiplied by 100. All three quantities are quite constant from the center of the layer up to about 10 cm above (or below) the center of the layer. The edge of the layer occurs at a position

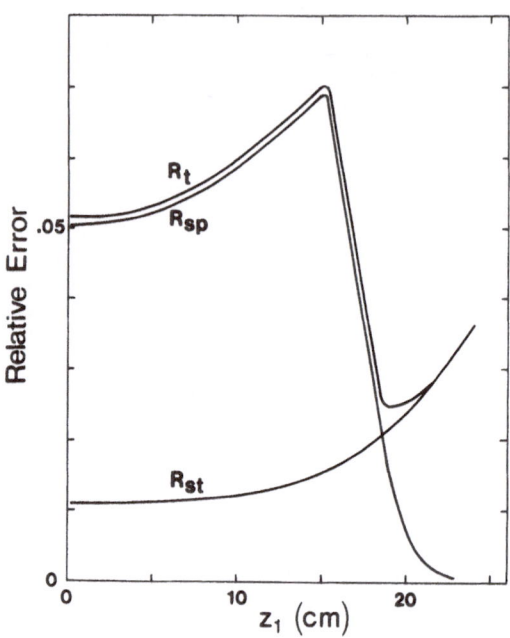

Figure 5.2.3 Total, statistical and spatial relative errors versus the axial position of a single measurement.

of 15 cm, and beyond about 10 cm the average response shows that the detector begins to "feel" the edge.

In figure 5.2.3 we show the total relative error of the measurement as well as the relative errors due to statistical and spatial uncertainties versus the axial position of the measurement. The relative errors increase as the measurement position moves from the center of the layer toward the edge. As the measurement moves outside the layer the spatial relative error falls rapidly while the statistical relative error rises ever more sharply. This defines a clear minimum in the total relative error several centimeters outside the uraniferous layer. This minimum relative error of 2.5% is appreciably better than the 5.2% minimum occurring at the center of the layer.

In figures 5.2.4 and 5.2.5 we examine the behavior of a two-measurement assay, where the position of the first measurement is

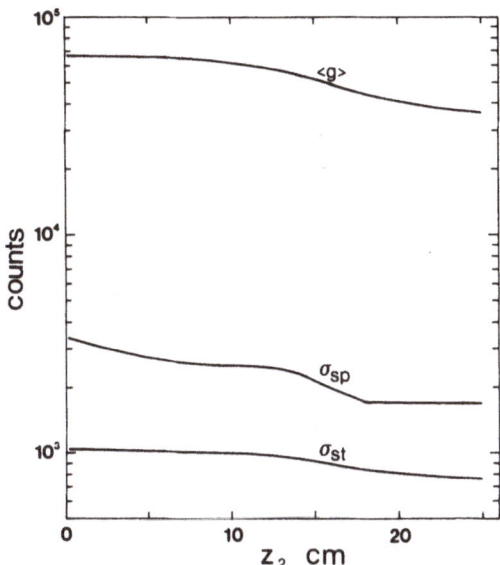

Figure 5.2.4 Average measurement and statistical and spatial standard deviations versus the axial position of the second of two measurements. The first measurement is performed at the center of the layer.

fixed at the center of the layer and the position of the second measurement varies. Comparison of figures 5.2.2 and 5.2.4 shows that the two-measurement case is generally more stable than the one-measurement situation. This effect is also evident in the relative errors shown in figure 5.2.5, which vary by only a small amount with the position of the second measurement. The least total relative error with two measurements is 4.3%, and occurs with the second measurement slightly outside the uraniferous layer. However, this relative error is greater than that obtained with a single measurement. This is because the spatial uncertainty never improves very much over the value at the center of the layer.

The dominating effect of the stationary center-of-layer measurement in the previous calculations suggests that both measurements should be allowed to move toward the periphery of the uraniferous layer. This can be done in many ways. In figures 5.2.6 and 5.2.7 we examine the behavior of two measurements separated by

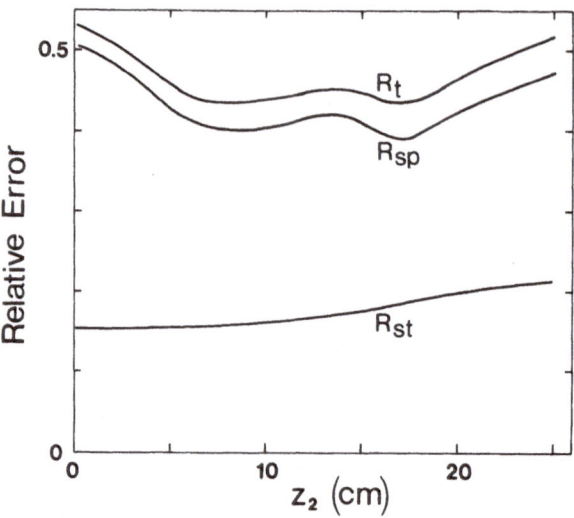

Figure 5.2.5 Total, statistical and spatial relative errors versus the axial position of the second of two measurements. The first measurement is performed at the center of the layer.

a constant distance of 10 cm. Comparison of these results with figures 5.2.2 and 5.2.3 shows qualitatively similar behavior. However an important difference is to be noted. Since one of the pair of measurements lags behind the other, the minimum in the total relative error occurs at a large distance outside the layer. As a consequence the statistical relative error is greater and the minimum in the total relative error is weaker.

Additional configurations can be examined. However, our conclusion up to now is that, on the basis of the total relative error, the best assay is obtained with a single measurement performed slightly outside the uraniferous layer. Also, the relative error of this measurement, being 2.5%, is slightly more than half the relative error of the next best measurement configuration studied.

There are several complicating factors which are not included in this formulation of the assay problem. Among these are variability of the mass of the uranium nodules and irregularity of the boundaries of the uraniferous layer. To the extent that data are

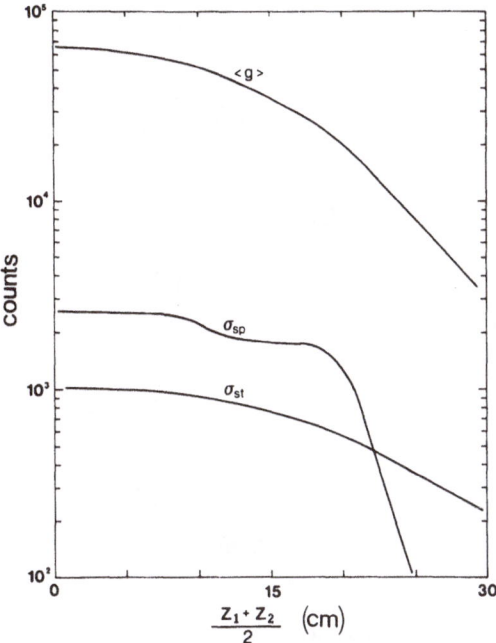

Figure 5.2.6 Average measurement and statistical and spatial standard deviations versus the midpoint between two measurements separated by 10 cm.

available to enable reasonable modelling of these factors, the analysis which we have performed can be readily extended to include them. It may be particularly important to assess the influence of boundary irregularities, since such irregularities may reduce the depth of the sharp minimum in the relative error occurring just outside the layer. It may also be important to assess the effect of variability of the absorption coefficient due to matrix heterogeneity.

5.3 MINIMUM VARIANCE CRITERION

5.3.1 Rao-Cramer Inequality

The design criterion which we discussed in the previous section

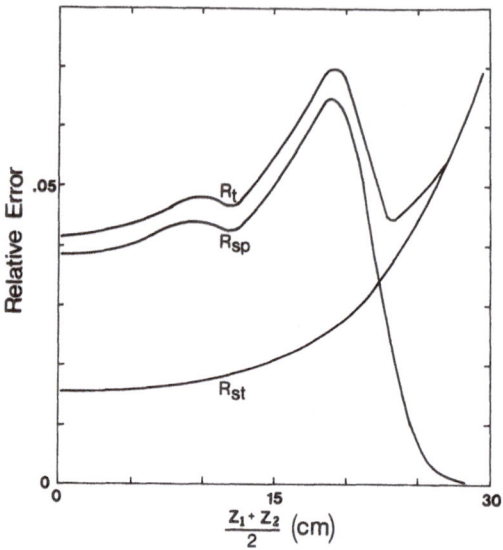

Figure 5.2.7 Total, statistical and spatial relative errors versus the mid-point between two measurements separated by 10 cm.

is based on the concept of relative error: the ratio of the standard deviation to the mean of some quantity. We showed in a detailed example how the relative error of the response may be evaluated without requiring knowledge of any probability density function. In the philosophy of relative-error design, the "best" system is the one for which the response varies by the least amount for a given quantity of analyte.

The minimum-variance design criterion is based on a different approach. The "best" design is the one for which it is possible to estimate, with least variance, the unknown parameter which is being assayed. While it is not necessary to construct or examine any particular estimation scheme for interpreting the measurements, it is necessary to know the conditional probability density of the response given the value of the assayed parameter. The fundamental tool for minimum variance design is the Rao-Cramer inequality, which we shall introduce and illustrate with a series of examples.

In order to discuss the Rao-Cramer inequality we must recall some standard terminology from statistical estimation theory.
We shall distinguish between *parameters*, whose values we wish
to estimate, and *random variables* which are measured and upon
which the estimation is based. Parameters which are commonly
estimated are the quantity, position or dimension of a material.
The random variables are the measured detector responses. A
statistic is a function of random variables which does not depend
explicitly on any unknown parameters. A statistic is itself a random variable. A statistic is *unbiased* for a given parameter if the
expectation of the statistic precisely equals the value of the parameter. An unbiased statistic is commonly called an unbiased
estimate.

Let $p(g|x)$ be the conditional probability density of the re-
sponse vector g given the parameter x (which may be the source
mass, average position, etc.). Let $g(1), \ldots, g(n)$ be n statistically
independent vector measurements. (Note that repeated measurements of the same sample will not generally be statistically independent with respect to the spatial variability of the source ma-
terial). Each of the $g(i)$ is distributed according to $p(g|x)$. Let
$y\big(g(1), \ldots, g(n)\big)$ be an unbiased statistic for the parameter x. The
Rao-Cramer inequality states that the variance of y satisfies [1]

$$\text{var}(y) \geq \left(n\, \mathrm{E}\left[\left(\frac{\partial \ln p(g|x)}{\partial x} \right)^2 \right] \right)^{-1} . \qquad (5.3.1)$$

where $\mathrm{E}(A)$ represents the expectation of the quantity A. We
notice that the right hand side of this relation does not depend
on the functional form of the statistic y. Thus this expression
constitutes a lower bound — the Rao-Cramer lower bound — on
the variance of *any* unbiased statistic for the parameter x. The
conditional density $p(g|x)$ embodies the design characteristics of
the assay system as well as the properties of the assay application.
Thus the Rao-Cramer lower bound specifies the best resolution
(in a minimum variance sense) which can be obtained, from the
assay-system in question, by any unbiased statistic. The Rao-

Cramer lower bound is thus a powerful probabilistic measure of performance which can be used for comparing alternative assay-system designs.

The following equation is readily proven, and in some cases provides easier computation of the Rao-Cramer lower bound than relation (1).

$$E\left[\left(\frac{\partial \ln p(g|x)}{\partial x}\right)^2\right] = -E\left[\frac{\partial^2 \ln p(g|x)}{\partial x^2}\right] \qquad (5.3.2)$$

The variance of an unbiased statistic y for the parameter x may be given a simple physical meaning. Since y is unbiased, its mean equals x. Thus the variance of y is

$$\mathrm{var}(y) = E\left[(y-x)^2\right] \quad .$$

In other words, the variance of y is the mean square error of the estimate.

In some situations more than one parameter is to be estimated, and each measurement consists of several correlated random variables. Let $x = (x_1, \ldots, x_n)$ be the parameter vector and let $g = (g_1, \ldots, g_n)$ be the measurement vector. The conditional probability density of the measurement vector given the parameter vector is $p(g|x)$. Let us define the matrix Q whose elements are

$$q_{ij} = E\left(\frac{\partial \ln p(g|x)}{\partial x_i} \frac{\partial \ln p(g|x)}{\partial x_j}\right)$$

$$= -E\left(\frac{\partial^2 \ln p(g|x)}{\partial x_i \partial x_j}\right) \qquad (5.3.3)$$

Q is known as Fisher's information matrix. Let $y_i(g)$ be an unbiased estimate for parameter x_i, based on the measurement vector g. The variance of $y_i(g)$ satisfies the following inequality:

$$\mathrm{var}(y_i) \geq q'_{ii} \qquad (5.3.4)$$

where q'_{ii} is the i-th diagonal element of the inverse of Q.

5.3.2 Examples

Many applications of the Rao-Cramer inequality may be found in the literature. In this section we shall discuss a few brief examples.

1. Single measurement of a normal distribution. Let the measurement g be a normally distributed scalar variable with conditional density

$$p(g|x) = \frac{1}{\sqrt{2\pi v}} \exp\left(-\frac{(g - bx)^2}{2v}\right) \quad .$$

Let $y(g)$ be an unbiased statistic for x. The Rao-Cramer inequality provides a lower bound for the variance of y. The logarithm of the conditional density is readily differentiated and its expectation found. The Rao-Cramer lower bound is

$$\mathrm{var}_{rc} = \frac{v}{b^2} \quad .$$

This simple result yields two points of interest. First, the designer should strive to make the variance, v, of the measurement as small as possible, since the minimum variance of the estimator of x increases linearly with v. Second, the designer should make the average response increase strongly with x. That is, he should strive to make b large. In many circumstances the parameters b and v are interrelated and can not be independently altered by design modifications. A design-analysis such as this precisely quantifies — in a minimum variance sense — the resolution capability of any given design.

2. Utility of the marginal measurement interval. Let the conditional probability density for the response g, given the parameter x, be $p(g|x)$. Let us perform n independent but identically distributed measurements in n disjoint intervals of time of uniform duration. The Rao-Cramer lower bound for the variance of an unbiased estimator of x is given by relation (1). The ratio of the minimum variance with n measurements to the minimum variance with one measurement is

$$\frac{\text{var}_{rc}(n)}{\text{var}_{rc}(1)} = \frac{1}{n} \qquad .$$

Thus we must double the number of measurement intervals in order to reduce the minimum variance by a factor of two. Alternatively we may express the utility of the marginal (n-th) measurement as the fractional change in the minimal variance:

$$\frac{\text{var}_{rc}(n-1) - \text{var}_{rc}(n)}{\text{var}_{rc}(n-1)} = \frac{1}{n} \qquad .$$

More sophisticated tools are available for evaluating the utility of sequential measurements [2], and will be discussed in section 6.2.

3. Subdivision of a fixed measurement duration. Let the response g, given the parameter x, be normally distributed with mean btx and variance vt, where t is the duration of the measurement. Thus the average response is proportional to the measurement duration, t, and to the estimated parameter x, while the variance of the measurement is independent of x. Let us suppose that we have a total of T seconds in which to measure. We shall divide this interval into n equal sub-intervals, each of duration T/n seconds. Let us suppose that the n measurements are statistically independent.

We may readily find the Rao-Cramer lower bound to be

$$\text{var}_{rc} = \frac{v}{b^2 T} \qquad .$$

Thus the minimal variance of an unbiased estimator for x is independent of how we subdivide the total allotted time.

This conclusion is of course limited to the particular class of assay-systems which are represented by the conditional probability density which we have defined. If we let the response g be normally distributed with mean btx (as before) but with variance vtx, we find the Rao-Cramer lower bound to be

$$\text{var}_{rc} = \frac{2vx^2}{nv + 2b^2 xT} \qquad .$$

Thus the minimum variance decreases as the number of subdivisions of the total measurement duration increase. What is more, if the variance of each individual measurement is much greater than the square of the mean, then

$$\frac{vxT}{n} \gg \left(\frac{bxT}{n}\right)^2$$

which implies that

$$nv \gg b^2 xT \quad .$$

In this case the Rao-Cramer lower bound becomes

$$\text{var}_{rc} = \frac{2x^2}{n} \quad .$$

Thus the total time of measurement has no effect on the minimum variance.

From this example we see that the dependence of the Rao-Cramer lower bound on the conditional probability density of the response must be taken seriously: modelling errors in constructing $p(g|x)$ may lead to unrealistic design decisions.

4. Efficiency of an unbiased estimator. The ratio of the Rao-Cramer lower bound to the variance of an unbiased statistic is the *efficiency* of that statistic. As a final example we shall illustrate use of the Rao-Cramer inequality for evaluating the efficiency of a statistic.

Let the probability density of the response, g, given the parameter x, be a gamma density:

$$p(g|x) = \frac{ge^{-g/x}}{x^2} \quad , \quad g > 0 \quad .$$

The mean and variance of the response are

$$\langle g \rangle = 2x$$
$$\text{var}(g) = 2x^2 \quad .$$

As our unbiased statistic for estimating x we shall choose

$$y(g) = \frac{1}{2}g \qquad .$$

The variance of this statistic is

$$\mathrm{var}(y) = \frac{1}{2}\mathrm{var}(g) = x^2 \qquad .$$

The Rao-Cramer lower bound is readily found to be

$$\mathrm{var}_{rc} = \frac{1}{2}x^2 \qquad .$$

Thus the efficiency of this statistic is $1/2$.

5.4 PROBABILISTIC EXPANSION

5.4.1 Definition of the Overlap Function

The simplest deterministic measure of performance developed in Chapters 2 and 3 is based on the geometrical concept of the expansion of the complete response set, C. We have defined the expansion as the supremum of the set of real numbers y for which C and yC intersect. That is

$$e(C) = \sup\{y : C \cap yC \neq \emptyset\} \qquad (5.4.1)$$

We showed in Chapter 2 that the expansion precisely equals the relative resolution of the assay system (without consideration of statistical uncertainty). For a zero-moment assay, for instance, every spatial distribution in the sample of u grams of source material yields a set of measurements different from that obtained from every spatial distribution of v grams if and only if

$$v > zu \qquad (5.4.2)$$

where z is the expansion.

We soon came to understand that the expansion usually provides a rather conservative estimate of the resolution capability of the system. By this we mean that, when v is only slightly less than zu, the probability is likely to be quite low that v grams will be spatially distributed in the sample in such a way as to yield a set of measurements which could arise from u grams. This means that a certain assay-system may in practice provide adequate resolution, even though the deterministic expansion indicates otherwise. In light of this situation it is clearly desirable to develop the capability for evaluating the probability that v grams of source material may be distributed in such a way as to yield a measurement obtainable from u grams. In other words, we are seeking a probabilistic generalization of the concept of expansion. It may be recalled that the utility of the concept of expansion is limited to situations in which the class of complete response sets has the property of proportionality. We would like our probabilistic generalization of the expansion to be free of this requirement.

Let us consider an n-measurement assay system. Let g represent the measurement vector and $p(g|u)$ the conditional probability density of g given u grams of source material in the sample. The following *overlap function* expresses the probability that u grams of source material will be distributed in such a way as to give a measurement which could arise from v grams:

$$K(u, v) = \int p(g|u) H\big(p(g|v)\big) \, dg \qquad (5.4.3)$$

where

$$H(x) = \begin{cases} 0 & \text{if } x \leq 0 \\ 1 & \text{if } x > 0 \end{cases} \qquad (5.4.4)$$

The function $K(u, v)$ contains all the information we need for evaluating the probability that u may be mistaken for v grams of source material. $K(u, v)$ is shown schematically in figure 5.4.1 for an assay system whose relative mass resolution is z. If v/u is greater than the relative mass resolution, then there are no values of g which can arise from both u and v grams, and $K(u, v)$ equals

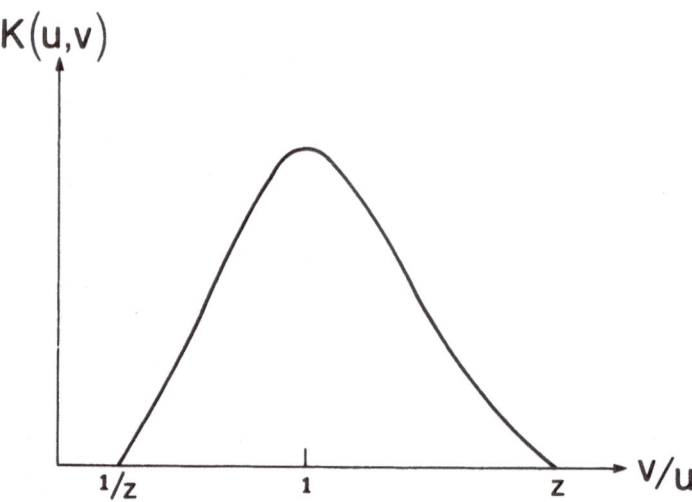

Figure 5.4.1 Schematic overlap function.

zero. As v/u decreases in value below z (for fixed u), $K(u,v)$ will tend to increase, until it finally reaches a value of unity at $v = u$. As v/u decreases to values less than unity $K(u,v)$ also decreases, reaching zero at (or before) $u/v = z$. Usually $K(u,v)$ is not symmetrical about $v = u$, the upper tail being longer. Furthermore, the shape of $K(u,v)$ versus v is likely to change with u.

The wealth of information contained in the dependence of $K(u,v)$ on both u and v makes the overlap function considerably less convenient as a design tool than the deterministic expansion z. However, once the overlap function is known for a range of values of u and v one may distill this information in a number of ways, for example by considering low-order moments of $K(u,v)$ rather than the entire function. On the other hand, one may wish to avoid having to construct the overlap function for a large range of values of u and v. This can be achieved if evaluation of $K(u,v)$ for selected u and v can be used to infer the values of $K(r,s)$ for a range of r and s. If this is possible one can reduce the probabilistic evaluation of the design to consideration of a limited range of values of u and v. Relations between $K(r,s)$ and $K(u,v)$ are thus known as *reduction theorems*.

In section 5.4.3 we shall introduce the subject of reduction theorems by considering some elementary inequalities for the overlap function, under somewhat restrictive conditions. Our aim in this discussion is two-fold: to provide some basic results of use to the designer, and to indicate a challenging and interesting field of investigation for the theorist. However, before discussing reduction theorems we shall examine a simple example of the overlap function and its use.

5.4.2 Example: Overlap Functions

The overlap function $K(u, v)$ expresses the probability that u grams of analyte material will distribute itself in such a way as to yield a response which could be obtained from v grams. The overlap function is evaluated from the conditional probability density of the response given the quantity of analyte, $p(g|u)$. This conditional density is, in turn, derived from the characteristics of the assay-system design and from the properties of the assay application.

Examples in Chapter 4 showed that, as u increases, $p(g|u)$ generally becomes narrower and its centroid increases. The detailed shape of the conditional probability density must of course be established for each application. However the following conditional density is somewhat typical, and enables ready evaluation of the overlap function. Let the conditional density be

$$p(g|u) = \frac{u+1}{2} \left(1 - |g - \langle g \rangle - uD|^{1/u} \right) \ , \ |g - \langle g \rangle - uD| \le 1$$

In figure 5.4.2 we illustrate this conditional density for $\langle g \rangle = 9.5$, $D = 0.5$ and several values of u.

The overlap function for this conditional density may be conveniently expressed in terms of the conditional probability distribution as

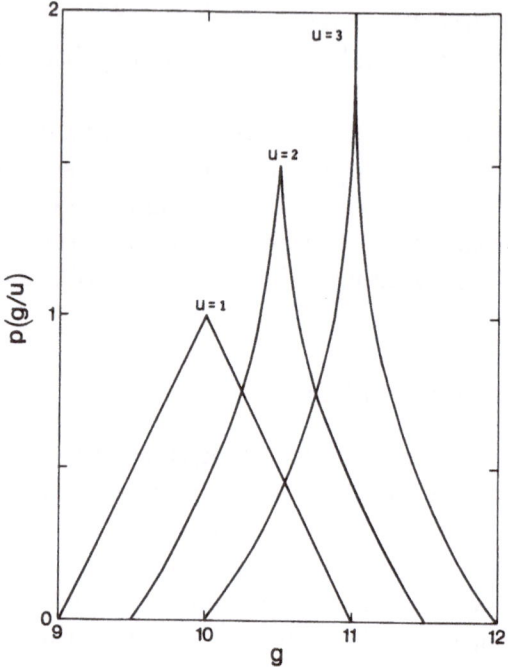

Figure 5.4.2 Conditional probability densities.

$$K(u,v) = \begin{cases} 1 - P(\langle g \rangle + vD - 1|u) & \text{if } u \le v \\ P(\langle g \rangle + vD + 1|u) & \text{if } u \ge v \end{cases}$$

The conditional distribution is found to be

$$P(x|u) = \frac{1}{2} - \frac{u+1}{2}(\langle g \rangle + uD - x) + \frac{u}{2}(\langle g \rangle + uD - x)^{(u+1)/u}$$

for $x \le \langle g \rangle + uD$. Since $p(g|u)$ is symmetrical about its mean, $\langle g \rangle + uD$, we have

$$P(x|u) = 1 - P\big(2(\langle g \rangle - uD) - x|u\big) \quad \text{for} \quad x \ge \langle g \rangle + uD \quad .$$

In figure 5.4.3 we show the overlap function $K(u,v)$ versus v for $u = 1$. The overlap vanishes for $v > 5$ since the domains of non-zero probability do not overlap. Furthermore we see that the probability is 0.13 that 1 unit of analyte will be spatially distributed so

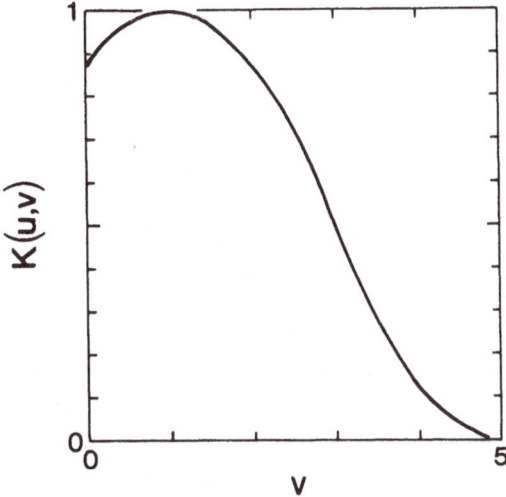

Figure 5.4.3 Overlap function $K(u,v)$ versus v, for $u=1$.

as to yield a response which could be obtained from 4 units; the probability is 0.50 that 1 unit of analyte will "imitate" 3 units; and the probability is 0.87 that 1 unit will imitate 2 units of analyte. By evaluating $K(u,v)$ versus v for various values of u we obtain a comprehensive probabilistic assessment of the resolution capability of the assay system. Comparison of these curves for alternative designs enables identification of the best design.

5.4.3 Reduction Theorems: Inequalities for the Overlap Function

Let us consider a single-detector system for the assay of identical source particles. The response to a single source particle ranges from g_1 to g_2 where $0 \leq g_1 \leq g_2$. Thus the response to m source particles, assuming linearity in mass, ranges from mg_1 to mg_2. Let us furthermore assume that the probability density of the response g, conditioned by the number of source particles m, exists and is zero on at most a countable number of points in $[mg_1, mg_2]$. We shall represent the conditional density and distribution by $p(g|m)$ and $P(g|m)$ respectively. Referring to eq.(3) and choosing $m \leq n$, we may express the overlap function as

$$K(m,n) = \int\limits_{ng_1}^{mg_2} p(g|m)\, dg = 1 - P(ng_1|m) \qquad (5.4.5)$$

and

$$K(n,m) = \int\limits_{ng_1}^{mg_2} p(g|n)\, dg = P(mg_2|n) \qquad (5.4.6)$$

From these relations and the fact that the probability distribution is a non-decreasing function we obtain the first most elementary inequality. (More powerful and general relationships can be developed which do not depend on the simplifying assumptions we have made [3, 4]. However it is beyond the scope of this book to discuss them).

THEOREM 1 Given the conditions specified above,

$$K(m,n) \geq K(m,n') \quad \text{if} \quad m < n < n' \qquad (5.4.7)$$

and

$$K(n,m) \geq K(n,m') \quad \text{if} \quad n > m > m' \qquad (5.4.8)$$

with equality if and only if

$$mg_2 \leq ng_1 \qquad (5.4.9)$$

in which case $K(m,n) = 0$.

PROOF Since $p(g|m)$ is zero on at most a countable number of points in $[mg_1, mg_2]$, $P(g|m)$ is strictly increasing on the same interval and of course constant elsewhere. The Theorem thus results directly from eqs. (5) and (6). QED

Eq.(7) states that the probability of m particles being indistinguishable from $n > m$ particles, is an upper bound for the probability of m and $n' > n$ particles being indistinguishable. Eq.(8) may be similarly interpreted. In other words, (for the class of assay problems which we are considering) the plot of $K(m,n)$ versus

n for fixed m displays steadily decreasing tails, above and below the point $m = n$, as shown in figure 5.4.1.

Theorem 1 allows comparison of different points on the same tail of $K(m, n)$ — either the upper or lower tail. It is now reasonable to enquire into the comparison of the two tails. We shall be satisfied to establish a relationship between $K(m, n)$ and $K(n, m)$. From eqs.(5) and (6) we see that, for $m < n$, $K(m, n)$ is derived from the upper tail of the probability density $p(g|m)$, while $K(n, m)$ is derived from the lower tail of the density $p(g|n)$. In order to relate $K(m, n)$ and $K(n, m)$ we must establish the following preliminary results.

LEMMA 1 Let $x_1, x_2, \ldots$ be identical random variables independently distributed on the interval $[g_1, g_2]$. Let the density and distribution of the sum of m such random variables exist and be denoted $p(x|m)$ and $P(x|m)$ respectively. Then, for $n > m$

$$P(ng_1 + t|n) \le P(mg_1 + t|m) \quad .$$

for all t. Equality is obtained for a given t if and only if $P(x|m)$ is constant on the interval $[ng_1 + t - (n - m)g_2, mg_1 + t]$.

PROOF Since the random variables are independent, $P(y|n)$ and $P(y|m)$ are related by the convolution

$$P(y|n) = \int_{y-(n-m)g_2}^{y-(n-m)g_1} P(x|m)p(y - x|n - m)\, dx$$

It is readily shown that this integral does not exceed the greatest value of $P(x|m)$ on the interval of integration. Since $P(x|m)$ is a non-decreasing function, we find that

$$P(y|n) \le P\big(y - (n - m)g_1|m\big)$$

Choosing $y = ng_1 + t$ yields the desired result. The sufficiency of the condition for equality is obvious. The necessity arises from the fact that $P(x|m)$ is continuous and non-decreasing. QED

DEFINITION 1 Let x be a random variable on the interval $[g_1, g_2]$ with distribution $P(x)$. We say that x is *sub-symmetrically distributed* if

$$P(g_1 + t) \leq 1 - P(g_2 - t) \tag{5.4.10}$$

for all t.

A special class of sub-symmetrical distributions of particular importance arises when the inequality in relation (10) is replaced by equality, yielding symmetrical distributions.

LEMMA 2 Let $x_1, x_2, \ldots$ be independent sub-symmetrical random variables identically distributed on $[g_1, g_2]$, where $0 \leq g_1 \leq g_2$. For $n = 1, 2, \ldots$, let the density and distribution of the sum of n such random variables exist and be denoted $p(x|n)$ and $P(x|n)$, respectively. Then $x_1 + \cdots + x_n$ is sub-symmetrically distributed on $[ng_1, ng_2]$.

PROOF That $x_1 + \cdots + x_n$ is distributed on $[ng_1, ng_2]$ is evident. We shall prove the remainder of the Lemma by induction. Consider the case $n = 2$. Since x_1 and x_2 are independent

$$P(2g_1 + x|2) = \int P(2g_1 + x - u|1)p(u|1)\, du$$

and

$$P(2g_2 - x|2) = \int P(2g_2 - x - u|1)p(u|1)\, du$$

Thus

$$P(2g_1 + x|2) + P(2g_2 - x|2)$$
$$= \int \left(P(2g_1 + x - u|1) + P(2g_2 - x - u|1) \right) p(u|1)\, du$$

Since $g_1 \geq 0$, $p(u|1)$ is non-zero only for $u \geq 0$. Thus since $P(x|1)$ is a non-decreasing function

$$P(2g_2 - x - u|1) \leq P(2g_2 - x + u|1)$$

Hence, since $P(x|1)$ is sub-symmetric,

$$P(2g_1 + x|2) + P(2g_2 - x|2)$$
$$\leq \int \big(P(2g_1 + x - u|1) + P(2g_2 - x + u|1)\big)p(u|1)\,du$$
$$\leq \int p(u|1)\,du$$
$$\leq 1$$

which shows that $P(x|2)$ is sub-symmetric. A similar argument may be employed to complete the proof inductively. QED

THEOREM 2 Let $p(g|m)$ be the conditional probability density of a single-detector system for the assay of randomly and independently distributed identical source particles. Assume linearity in mass. Let the response to one particle be sub-symmetrically distributed on $[g_1, g_2]$, where $0 \leq g_1 \leq g_2$. Let $p(g|m)$ equal zero on at most a countable number of points in $[mg_1, mg_2]$. Then for $n > m$

$$K(n, m) \leq K(m, n)$$

with equality if and only if

$$mg_2 \leq ng_1$$

PROOF First of all, let us note that since $P(g|1)$ is sub-symmetrical, Lemma 2 implies that $P(g|m)$ is sub-symmetrical for all m. Recalling eq.(5) we have

$$K(m, n) = 1 - P\big(mg_2 - (mg_2 - ng_1)|m\big)$$
$$\geq P\big(mg_1 + (mg_2 - ng_1)|m\big) \qquad (5.4.11)$$

with equality if and only if

$$P(ng_1|m) + P(mg_2 + mg_1 - ng_1|m) = 1 \qquad (5.4.12)$$

From eq.(6) we have

$$K(n, m) = P\big(ng_1 + (mg_2 - ng_1)|n\big) \qquad (5.4.13)$$

From Lemma 1 we obtain

$$P\big(mg_1 + (mg_2 - ng_1)|m\big) \geq P\big(ng_1 + (mg_2 - ng_1)|n\big) \quad (5.4.14)$$

Strict equality is obtained if and only if $P(x|m)$ is constant on the interval
$$I = [mg_2 - (n - m)g_2, mg_2 - (n - m)g_1]$$

Since $p(g|m)$ is zero on at most a countable number of points in $[mg_1, mg_2]$, we conclude that $P(g|m)$ is strictly increasing on $[mg_1, mg_2]$ and of course constant elsewhere. Thus $P(z|m)$ is constant on the interval I if and only if

$$mg_2 - (n - m)g_1 \leq mg_1$$

which is equivalent to

$$mg_2 \leq ng_1 \qquad\qquad (5.4.15)$$

Combining relations (11), (13) and (14) yields

$$K(m, n) \geq K(n, m)$$

with equality if and only if relation (15) holds. QED

Roughly speaking, this Theorem states that it is less likely that a large number of particles will be indistinguishable from a small number, than that a small number will be indistinguishable from a large number. The most important condition limiting the generality of the Theorem is the supposition that g is distributed sub-symmetrically. However as we have noted, the property of sub-symmetry includes the important and widespread case of symmetrical distributions.

While Theorem 2 establishes sub-symmetry as a sufficient property for the indicated inequality, we can readily see that it is not a necessary condition. For $m < n$, let us define

$$E_{mn}(x) = P(mg_1 + x|m) - P(ng_1 + x|n) \qquad (5.4.16)$$

and

$$W_m(x) = 1 - P(mg_2 - x|m) - P(mg_1 + x|m) \qquad (5.4.17)$$

Thus E_{mn} compares the lower tails of $P(g|m)$ and $P(g|n)$, while W_m compares the lower and upper tails of $P(g|m)$. Lemma 1 asserts that E_{mn} is always non-negative, and in fact positive if $P(g|m)$ is strictly increasing. On the other hand W_m is non-negative only for sub-symmetrical distributions. Adding eqs.(16) and (17) yields

$$E_{mn}(x) + W_m(x) = 1 - P(mg_2 - x|m) - P(ng_1 + x|n) \quad (5.4.18)$$

Upon choosing

$$x = mg_2 - ng_1 \qquad (5.4.19)$$

eq.(18) becomes

$$E_{mn}(mg_2 - ng_1) + W_m(mg_2 - ng_1) = K(m, n) - K(n, m) \quad (5.4.20)$$

Thus the inequality of Theorem 2 can be maintained even if $P(g|m)$ violates sub-symmetry, provided the left-hand side of eq.(20) remains non-negative.

NOTES

[1] This inequality was first formulated by R. A. Fisher in
1. R. A. Fisher, On the Mathematical Foundations of Theoretical Statistics, *Proc. Roy. Soc.*, London, 222: 309 (1922).
However, this Theorem is usually associated with the work of Rao and Cramer:
2. R. Rao, Information and Accuracy Attainable in the Estimation of Statistical Parameters, *Bull. Calcutta Math. Soc.*, 37: 81-91 (1945).
3. H. Cramer, *Mathematical Methods of Statistics*, Princeton Univ. Press, 1946.
The Theorem is discussed in many standard texts, including

4. R. V. Hogg and A. T. Craig, *Introduction to Mathematical Statistics*, Macmillan, 1970.

[2] For a general discussion of sequential analysis see
1. H. Chernoff, *Sequential Analysis and Optimal Design*, S.I.A.M. monograph, 1972.
Applications of sequential analysis in the assay of nuclear materials are discussed in
2. P. E. Fehlau, K. L. Coop and K. V. Nixon, Sequential Probability Ratio Controllers for Safeguards Radiation Monitors, 6-th ESARDA Symp. on Safeguards and Nuclear Mat. Mgt., pp 155-7, Venice, May 1984.
3. P. E. Fehlau, K. L. Coop and J. T. Markin, Applications of Wald's Sequential Probability Ratio Test to Nuclear Materials Control, ESARDA Specialist's Working Group on Statistical Problems in Nondestructive Assay, Ispra, Italy, Sept. 1984.

[3] Y. L. Tong, *Probability Inequalities in Multivariate Distributions*, Academic Press, 1980.

[4] Y. Ben-Haim, Convex Sets and Nondestructive Assay, *S. I. A. M. J. Algebraic and Discrete Methods*, to appear.

CHAPTER 6

ADAPTIVE ASSAY

6.1 MOTIVATION

The assay of material is comprised of four fundamental tasks: modelling, design, measurement and interpretation. These may be arranged as in figure 6.1.1. The model of the sample must provide all available information pertinent to the design process. The sample model constitutes the input upon which the design analysis is performed. The optimization of an assay-system design is based on a measure of performance: an index which quantitatively evaluates the performance of a proposed design, in the face of defined spatial, statistical or other uncertainties. For a given definition of the assay problem, including precise statement of the uncertainties involved, the measure of performance enables rational selection of the assay-system design. Data interpretation is based on a decision rule whereby information about the sample is extracted from measurements obtained by the assay system. The past four Chapters have been devoted to a thorough study of design-optimization and data-interpretation.

Model of Sample	Design Optim- ization		Measurement		Data Interpre- tation
	$\longrightarrow$		$\longrightarrow$	$\longrightarrow$	

Figure 6.1.1 Information flow diagram for the assay of material.

Formulation of the sample model is usually not a trivial task. An extensive data base and elaborate mathematical modelling may be required in order to accurately represent the full complexity and range of variability of the samples. In the absence of adequate formulation of the model, the best design analysis may be of limited value. Alternatively, if the range of variability of the samples is great, then even the most realistic model will not facilitate a conclusive optimization of design: the overall optimal design is apt to be either extraordinarily complex in order to account for the full range of variability of the samples, or to be far from optimal for many individual specimens.

Extensive prior knowledge of each individual specimen, as well as the time to exploit such knowledge, are often lacking in real assay situations. However a simple and natural modification of the assay scheme of figure 6.1.1 can extricate the analyst from these difficulties. The *adaptive assay* scheme shown in figure 6.1.2 greatly extends the assay power of simple instrumentation and is indispensible when confronting uncertainty in the definition of the sample.

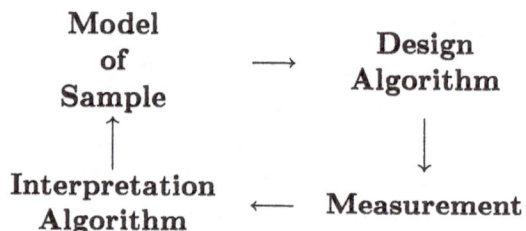

Figure 6.1.2 Information flow diagram for adaptive assay.

The basic idea of adaptive assay is that the assay system design is adjusted in the course of the measurement so as to optimize the assay. The assay system is operated under computer control, and the computer is programmed to interpret the results of measurement on-line, and to adapt the assay-system configuration in accordance with the updated understanding of the sample.

An adaptive assay is comprised of the four components shown in figure 6.1.2:

1. A model of the sample and of the detectors. This model must explicitly specify those model-parameters which influence the accuracy of the assay, and those detector-parameters which are subject to design modification.

2. A design algorithm by which the design is optimized. This algorithm must define a measure of performance by which the optimal detector parameters are specified, given estimated values of the model parameters.

3. A system for performing and recording measurements of the sample.

4. An interpretation algorithm for estimating the parameters of the sample-model on the basis of the measurements.

Adaptive assay can take on various degrees of sophistication. In this Chapter only simple formulations will be introduced. In Chapter 7 some more advanced concepts will be briefly discussed. In section 6.2 the classical statistical technique of sequential analysis is described and presented as a tool of adaptive assay. By employing a sequential likelihood ratio test the duration of measurement can be determined in the course of measurement, thereby yielding appreciable savings in time. In sections 6.3 to 6.5 three different measures of performance are employed in adaptive assays in which the assay configuration is modified during measurement. In section 6.3 the relative mass resolution is employed as the measure of performance in the design algorithm, in section 6.4 the design optimization is based on a minimum variance criterion, and in section 6.5 the concept of relative error is utilized.

6.2 SEQUENTIAL ANALYSIS

6.2.1 Formulation

A design-parameter of importance in many applications is the duration of measurement. In the simplest adaptive assay problems, the duration of measurement is determined during the assay itself by applying a statistical test to the data as they are obtained. The assay is performed as a sequence of distinct and statistically independent measurements. After each measurement the data are analyzed and a decision is made whether to terminate the assay or to perform an additional measurement. This technique of *sequential analysis* has been extensively studied [1], and in this section our aim is to present a brief review of its basic features.

To introduce the technique, let us consider a measurement, g, on the basis of which we wish to distinguish between two different conditions in the sample. For example, in the case of quantitative assay, we may wish to determine whether or not the analyte concentration exceeds a threshold value; in fluid flow measurements we may wish to distinguish between alternative flow regimes. To use the statistician's terminology, one of the conditions is referred to as the null hypothesis H_0, and the other condition is the alternative hypothesis H_1. The conditional probability density of the measurement, given that hypothesis H_i is correct, is $p(g|i)$. The set of all possible responses (under either hypothesis) is G. We shall divide this complete response set into two decision regions G_0 and G_1, where G_i is the set of responses which are to be intepreted as arising under hypothesis H_i.

The probability of incorrectly interpreting hypothesis H_i is

$$P_i = \int_{G_j} p(g|i)\, dg \qquad j \neq i \quad , \quad i = 0 \text{ or } 1 \tag{6.2.1}$$

In section 4.4 we showed how to choose the decision regions so as to minimize P_1 while constraining P_0 to equal a constant c less than unity. We found that the decision regions are determined by the probability ratio (or likelihood ratio). Explicitly,

$$g \in G_0$$

if and only if

$$L(g) = \frac{p(g|1)}{p(g|0)} < x \qquad . \tag{6.2.2}$$

The critical value x of the probability ratio is related to the constant c by

$$c = \int_x^{\infty} p(L|0)\, dL$$

where $p(L|0)$ is the probability density of the probability ratio, under hypothesis H_0.

Eq.(2) defines a decision rule given a measurement g: if $L(g) < x$, then decide on hypothesis H_0, while if $L(g) \geq x$, then choose H_1. When the probability ratio is much less than (or much greater than) the critical value, the decision is reasonable and likely to be correct. However, when $L(g)$ is near x the measurement is ambiguous. The technique of sequential analysis reduces this ambiguity, as we shall now explain.

Consider a sequence of independent identically distributed measurements represented by $g_1, \ldots, g_J$. The conditional probability density of this set of measurements, under hypothesis H_i, is

$$p(g_1, \ldots, g_J|i) = \prod_j p(g_j|i) \qquad (6.2.3)$$

The probability ratio for these measurements is

$$L(g_1, \ldots, g_J) = \frac{p(g_1, \ldots, g_J|1)}{p(g_1, \ldots, g_J|0)} \qquad (6.2.4)$$

Let x and y be positive numbers such that $x < y$. Our decision rule is as follows.

If $\quad L(g_1, \ldots, g_J) < x \quad$, $\quad$ then decide H_0

If $\quad L(g_1, \ldots, g_J) > y \quad$, $\quad$ then decide H_1

If $\quad x \leq L(g_1, \ldots, g_J) \leq y \quad$, $\quad$ perform another measurement

As before, the critical values x and y are related to the probabilities of incorrect decision. With this sequential approach we defer making a decision until the choice is clear cut to a certain extent.

It can be shown that this sequential procedure always terminates after a finite number of steps. What is more, it achieves a minimum average sample size (number of measurements) for given probabilities of error. This technique can be generalized to more complex situations, in which more than two hypotheses are involved. Discussion of such applications can be found elsewhere [2].

We shall illustrate this sequential decision rule with a few simple examples. However, before doing so it is important to discuss the relation between the critical values x and y and the probabilities of error.

Consider a sequential assay which terminates according to the above decision rule. Let $G_0(n)$ be the set of all n-tuple measurements which cause termination immediately after the n-th measurement (and not before), and which lead to choosing H_0. That is, $G_0(\mathrm{n})$ is the set of n-tuple measurements such that

$$x \leq L(g_1, \ldots, g_k) \leq y \quad , \quad k = 1, 2, \ldots, n-1$$
$$L(g_1, \ldots, g_n) < x \quad .$$

Likewise, let $G_1(n)$ be the set of n sequential measurements which terminate in deciding H_1. Thus $G_1(n)$ is the set of n-tuples such that

$$x \leq L(g_1, \ldots, g_k) \leq y \quad , \quad k = 1, 2, \ldots, n-1$$
$$L(g_1, \ldots, g_n) > y \quad .$$

Eq.(1) expresses the probability of error on the basis of a single measurement. We now wish to evaluate the probability that a sequential analysis will terminate in a false decision. The overall or sequential probability of incorrectly rejecting H_0 is

$$A = \sum_{n=1}^{\infty} \int_{G_1(n)} p(g_1, \ldots, g_n | 0) \, dg_1 \ldots dg_n \quad .$$

Similarly, the overall probability of incorrectly rejecting H_1 is

$$B = \sum_{n=1}^{\infty} \int_{G_0(n)} p(g_1, \ldots, g_n | 1) \, dg_1 \ldots dg_n \quad .$$

Since the probability of reaching a decision is unity, we have

$$1 - A = \sum_{n=1}^{\infty} \int_{G_0(n)} p(g_1, \ldots, g_n | 0) \, dg_1 \ldots dg_n$$

and

$$1 - B = \sum_{n=1}^{\infty} \int_{G_1(n)} p(g_1, \ldots, g_n | 1) \, dg_1 \ldots dg_n \quad .$$

Now, for each n-tuple in $G_1(n)$ we have

$$p(g_1, \ldots, g_n | 1) > y \, p(g_1, \ldots, g_n | 0) \quad .$$

Thus

$$A < \sum_{n=1}^{\infty} \int_{G_1(n)} \frac{1}{y} p(g_1, \ldots, g_n | 1) \, dg_1 \ldots dg_n$$

$$= \frac{1}{y}(1 - B) \tag{6.2.5}$$

Similarly, for each n-tuple in $G_0(n)$ we have

$$p(g_1, \ldots, g_n | 1) < x \, p(g_1, \ldots, g_n | 0) \quad .$$

Thus

$$B < \sum_{n=1}^{\infty} \int_{G_0(n)} x \, p(g_1, \ldots, g_n | 0) \, dg_1 \ldots dg_n$$

$$= x(1 - A) \tag{6.2.6}$$

From relations (5) and (6) we obtain the following inequalities relating the sequential error probabilities and the critical values:

$$\frac{A}{1 - B} < \frac{1}{y} \quad , \quad \frac{B}{1 - A} < x \tag{6.2.7}$$

These inequalities do not allow explicit determination of the critical values x and y in terms of the error probabilities A and B. However, the following approximation is frequently reliable. Let a and b be proper fractions such that

$$x = \frac{b}{1-a} \quad , \quad y = \frac{1-b}{a} \quad . \tag{6.2.8}$$

Then the error probabilities are approximately related to the quantities a and b by

$$A \approx a \quad , \quad B \approx b \quad . \tag{6.2.9}$$

6.2.2 Example: Batch Assay

Let us consider a collection of samples, all of which are generated by the same process. We are to perform a single measurement on each of a sequence of samples, in order to determine whether or not a certain analyte is present in the samples. If no analyte is present in the samples the probability density of each measurement is $p(g|0)$, while if analyte is present the probability density of each response if $p(g|1)$. If J different samples are measured, then the J measurements are independent and identically distributed random variables. The probability density of these J measurements, assuming hypothesis H_i ($i = 0$ or 1), is expressed by eq.(3).

Let us suppose that each measurement, g, has a gamma distribution. Explicitly,

$$p(g|0) = \frac{g}{a^2} e^{-g/a}$$

$$p(g|1) = \frac{g}{b^2} e^{-g/b}$$

where $b > a > 0$. Thus the probability ratio for J measurements is

$$L(g_1, \ldots, g_J) = \left(\frac{a}{b}\right)^{2J} \exp\left(\frac{b-a}{ba} \sum_1^J g_j\right)$$

Let x and y be the critical values, and define

$$X(J) = \frac{ab}{b-a}\left(2J\ln\frac{b}{a} + \ln x\right)$$

$$Y(J) = \frac{ab}{b-a}\left(2J\ln\frac{b}{a} + \ln y\right)$$

$$h(J) = \sum_1^J g_j \quad .$$

After some manipulations the decision rule becomes:

If	$h(J) < X(J)$	,	then choose H_0
If	$h(J) > Y(J)$	,	then choose H_1
If	$X(J) \le h(J) \le Y(J)$	,	no decision

We can calculate the error probabilities as follows. The random variable $h(J)$ is the sum of J independent random variables with a common gamma distribution. If hypothesis H_0 holds (no analyte present), the probability density of $h(J)$ is

$$p\big(h(J)|0\big) = \frac{h^{2J-1}}{(2J-1)!\,a^{2J}}\,e^{-h/a}$$

If hypothesis H_1 holds, then

$$p\big(h(J)|1\big) = \frac{h^{2J-1}}{(2J-1)!\,b^{2J}}\,e^{-h/b}$$

It is now easy to calculate the error probabilities given J measurements. The probability of making an incorrect decision when hypothesis H_0 holds is

$$P_0(J) = \int_{Y(J)}^{\infty} p(h|0)\,dh \quad .$$

Likewise the probability of falsely choosing H_0 on the basis of J measurements is

$$P_1(J) = \int\limits_0^{X(J)} p(h|1)\, dh \quad .$$

In a similar fashion we can express the probability of correctly identifying H_0 from J measurements as

$$D_0(J) = \int\limits_0^{X(J)} p(h|0)\, dh \quad ,$$

while the probability of correctly identifying H_1 is

$$D_1(J) = \int\limits_{Y(J)}^{\infty} p(h|1)\, dh \quad .$$

These probabilities can be succinctly expressed in terms of the tail of a χ^2 distribution. Let

$$Q(z|2m) = \frac{1}{2^m(m-1)!} \int\limits_z^{\infty} t^{m-1} e^{-t/2}\, dt \quad .$$

Then we obtain

$$P_0(J) = Q\left(\frac{2Y(J)}{a}\bigg|4J\right)$$

$$P_1(J) = 1 - Q\left(\frac{2X(J)}{b}\bigg|4J\right)$$

$$D_0(J) = 1 - Q\left(\frac{2X(J)}{a}\bigg|4J\right)$$

$$D_1(J) = Q\left(\frac{2Y(J)}{b}\bigg|4J\right)$$

The probability of not making any decision on the basis of J measurements, given hypothesis H_i, is

$$U_i(J) = 1 - P_i(J) - D_i(J) \qquad , \qquad i = 0 \text{ or } 1 \quad .$$

We shall consider a numerical example. Let $a = 1$, $b = 2$, $x = 1/3$ and $y = 7$. The following decision probabilities result.

J	P_0	P_1	D_0	D_1	U_0	U_1
1	0.00730	0.00005	0.0132	0.0992	0.980	0.901
2	0.0134	0.0725	0.312	0.220	0.675	0.708
3	0.0170	0.0875	0.462	0.321	0.521	0.592
4	0.0186	0.0870	0.561	0.412	0.420	0.501
5	0.0192	0.0808	0.631	0.483	0.350	0.436
6	0.0193	0.0764	0.686	0.550	0.295	0.374
7	0.0184	0.0692	0.729	0.606	0.253	0.325

This table indicates that the probability of not reaching a decision on the basis of a single measurement is quite large. Thus the error probabilities P_i and the correct-decision probabilities D_i are small for a single measurement. As the number of measurements increases, the probability of no decision falls while the probability of reaching a correct decision rises.

6.2.3 Example: Flow Rate Measurement

In this example we shall formulate a sequential analysis for determining the flow rate of a fluid. The fluid is activated (or a tracer is injected) at a fixed point along the flow channel. The activity of the fluid is measured during N disjoint time intervals after activation of the fluid. The measurement is performed by a colli-mated detector positioned at a known distance downstream from the activation site. Our aim is to distinguish between two distinct values which the flow rate may assume. Hypotheses H_0 and H_1 represent these alternatives. The average number of counts which would be recorded in the n-th time interval, assuming hypothesis H_i to hold, is $a_n(i)$. This function is known.

In the interest of economy and radiation safety, a fairly weak radiation source is used for activating the fluid. Thus the measured activity is low and shows considerable statistical fluctuations. We shall assume that the number of counts recorded in the n-th time interval is a Poisson random variable. Thus the probability distribution for the actual number of counts c_n in the n-th interval, assuming hypothesis H_i, is

$$p(c_n|i) = \frac{a_n(i)^{c_n} e^{-a_n(i)}}{c_n!}$$

The composite probability distribution $p(c_1, \ldots, c_N|i)$ for the N measurements performed after each activation, $c_1, \ldots, c_N$ may be evaluated as in eq.(3).

Having activated the fluid and made N measurements, we wish to decide whether to make an additional activation and perform an additional set of N measurements, or whether to choose between the two alternative hypotheses. In the case of an additional acti-vation, it is assumed that the effect of the previous activation has vanished. More generally, suppose we have made J activations, each followed by measurements performed during N time inter-vals. Let c_{nm} be the measured activity in the n-th interval after the m-th activation. The probability ratio for J sets of N-fold measurements is

$$L(J) = \prod_{m=1}^{J} \frac{p(c_{1m}, \ldots, c_{Nm}|1)}{p(c_{1m}, \ldots, c_{Nm}|0)}$$

After some manipulation one finds that

$$\ln L(J) = JW + \sum_{n=1}^{N} s_n \ln u_n$$

where

$$W = \sum_{n=1}^{N} \left(a_n(0) - a_n(1)\right)$$

$$u_n = \frac{a_n(1)}{a_n(0)}$$

$$s_n = \sum_{m=1}^{J} c_{nm}$$

Thus W and the u_n are known constants, and s_n is the sum of all the measurements obtained in the n-th interval after each activation.

Letting x and y be the critical values as before, the decision rule becomes

If $\displaystyle\sum_{n=1}^{N} s_n \ln u_n < \ln x - JW$,   then choose H_0

If $\displaystyle\sum_{n=1}^{N} s_n \ln u_n > \ln y - JW$,   then choose H_1

If $\ln x - JW \le \displaystyle\sum_{n=1}^{N} s_n \ln u_n \le \ln y - JW$,   no decision

This analysis yields an explicit expression for the statistic on which a decision is to be based. This statistic is particularly interesting since it shows that the results of the individual activations need not be stored. Rather, it is sufficient to accumulate the number of counts in each measurement interval. Regardless of how many activations have been performed, only the N numbers $s_1, \ldots, s_N$ need be retained and continually updated.

6.3 ADAPTIVE BARREL ASSAY

This section is devoted to the first example of an adaptive assay in which the detector configuration, rather than only the duration of measurement, is modified in the course of measurement. We shall illustrate the use of the relative-resolution as a measure of performance. A simple example will suffice to demonstrate the major features of this approach to adaptive assay.

6.3.1 The Model

Let the analyte material be arbitrarily distributed along a line segment of unit length. Thus the domain of the sample may be conveniently represented as $X = [0, 1]$. The sample-model is determined by a single parameter: the total mass of the analyte. The detector is positioned along the line defined by the sample, anywhere in the interval $Y = [2, \infty)$. The point-source response function is

$$f(x, y) = \frac{K}{(y - x)^2} e^{-\mu(1-x)}$$

where μ is the linear absorption coefficient of the sample matrix and K is a constant.

6.3.2 The Design Algorithm

The design algorithm must specify the position at which the detector is to be located. Our first task is to establish the measure of performance. This is an elementary matter for this single-measurement system. To find the relative mass resolution in the absence of statistical uncertainty, z, we construct the point-source response set F. This is the interval

$$F = [f(0, y), f(1, y)]$$
$$= \left[\frac{K e^{-\mu}}{y^2}, \frac{K}{(y - 1)^2} \right]$$

Since F is a convex set, we know that z, the relative mass resolution, is precisely the expansion of F. Thus

$$z = \left(\frac{y}{y-1} \right)^2 e^{\mu} \quad .$$

To evaluate the relative mass resolution in the presence of statistical uncertainty we must specify the duration of measurement t, the mass of analyte m, and the level of statistical confidence b. From eq.(2.11.3) we find

$$Z = \left(\sqrt{z} + \frac{b}{\sqrt{mtf(0,y)}} \right)^2$$

$$= e^{\mu} \left(\frac{y}{y-1} + \frac{by}{\sqrt{Kmt}} \right)^2 \quad .$$

We notice that the measure of performance Z is a function of the design parameter y, and of the sample-model parameter m. Thus given an estimate of m, we may find the value of y which minimizes Z. This optimal detector position is

$$\hat{y}(m,t) = 1 + \frac{(Kmt)^{1/4}}{\sqrt{b}} \quad .$$

If this expression is less than 2 then the optimum position is 2, in accordance with the constraint on the detector position. We have now completed our formulation of the design optimization rule.

6.3.3 The Interpretation Algorithm

In order to formulate an algorithm for interpretation of measurement we must have some information about the stochastic properties of the sample. The *a priori* probability density of the total mass of analyte is known to be

$$p(m) = \frac{1}{\sqrt{2\pi V}} \exp \left(-\frac{(m-M)^2}{2V} \right)$$

where M and V are known. Let $\langle g \rangle$ be the time-averaged number of counts obtained in t seconds with a detector at position y, from a sample containing m grams of analyte. The quantity $\langle g \rangle$ is a random variable because the spatial distribution of the analyte varies from sample to sample. The probability density of $\langle g \rangle$ is

$$ p_{sp}(\langle g \rangle | m; t, y) = \left(\frac{y^4}{2\pi t^2 W} \right)^{1/2} \exp \left(- \frac{\left(\langle g \rangle - \frac{tsm}{y^2} \right)^2}{\frac{2t^2 W}{y^4}} \right) . $$

So far we have only described the spatial uncertainty of the measurement; we must now consider the statistical uncertainty. Suppose we are measuring a particular sample whose time-average response is $\langle g \rangle$ in t seconds for a detector at point y. Then the probability density for an actual measurement of duration t at position y is

$$ p_{st}(g | \langle g \rangle; t, y) = \frac{1}{\sqrt{2\pi \langle g \rangle}} \exp \left(- \frac{(g - \langle g \rangle)^2}{2\langle g \rangle} \right) . $$

Now suppose we have obtained a set of N measurements from a given sample. Each measurement is of duration t, but at a different position; measurement g_i was taken from point y_i. We wish to estimate the total mass of analyte on the basis of these measurements. In section 4.3.7 we showed that this estimate is

$$ \widehat{m}(g_1, \ldots, g_N) = \int m p(m | g_1, \ldots, g_N) \, dm . $$

The conditional probability density in this integral is unknown. However, from the definition of conditional probability we have

$$ p(m | g_1, \ldots, g_N) = \frac{p(m, g_1, \ldots, g_N)}{p(g_1, \ldots, g_N)} $$

$$ = \frac{p(g_1, \ldots, g_N | m) \, p(m)}{\int p(g_1, \ldots, g_N | m) \, p(m) \, dm} . $$

Since the quantities $g_1, \ldots, g_N$ are actual measurements, they are distorted by both statistical and spatial uncertainty. Hence the conditional probability density in this last expression must contain both factors. If N is sufficiently large, we may approximate the true time-averaged value of the response with this sample, normalized to $t = y = 1$, as

$$\langle g \rangle = \frac{1}{N} \sum_{n=1}^{N} \frac{g_n y_n^2}{t} \quad .$$

Thus

$$p(g_1, \ldots, g_N | m) = p_{sp} \left(\langle g \rangle | m; t, y \right) \prod_{n=1}^{N} p_{st} \left(g_n \left| \frac{t \langle g \rangle}{y_n^2} ; t, y_n \right. \right) \quad .$$

Employing the above relations and the explicit forms for the probability densities we arrive at the following expression for the best estimate of the analyte mass. Given a set of measurements whose normalized average is $\langle g \rangle$, the best estimate of m is

$$\widehat{m} \left(\langle g \rangle \right) = \frac{\langle g \rangle / s + M \frac{W}{V s^2}}{1 + \frac{W}{V s^2}} \quad .$$

This expression can be given an interesting interpretation. M is the mean mass of analyte in the entire ensemble of samples. In other words, M is the *a priori* best estimate of the analyte mass in a given sample. The quantity $\langle g \rangle / s$ is the value of mass which is most likely to have given rise to the measured value $\langle g \rangle$. That is, $\langle g \rangle / s$ is the value of m which maximizes $p_{sp} \left(\langle g \rangle | m; t = y = 1 \right)$. Thus $\langle g \rangle / s$ is the *a posteriori* maximum likelihood estimate. Now the estimate $\widehat{m} \left(\langle g \rangle \right)$ is a weighted average of M and of $\langle g \rangle / s$. Explicitly,

$$\widehat{m} \left(\langle g \rangle \right) = \beta M + (1 - \beta) \frac{\langle g \rangle}{s}$$

where

$$\beta = \frac{\frac{W}{Vs^2}}{1 + \frac{W}{Vs^2}} \quad .$$

The weighting term β depends on the ratio of the variance of the conditional probability $p_{sp}(g|m;t,y)$ to the variance of the *a priori* density $p(m)$. If W is much larger than Vs^2 then the *a priori* estimate is likely to be more reliable than the *a posteriori* estimate. If W is much less than Vs^2, then the reverse is true. In any event, $\widehat{m}(\langle g \rangle)$ defines our rule for estimating model-parameters from measurements.

6.3.4 Implementation

We have a total measurement duration for each sample of T seconds, which we shall divide into N equal intervals of length t. We start the assay by choosing an *a priori* guess of the analyte mass, $m(1)$. It is reasonable to choose:

$$m(1) = M \quad .$$

From this estimate of the model-parameter, we evaluate the best detector position:

$$y(1) = \widehat{y}(m(1), T) \quad .$$

Now we position the detector and make a measurement, $g(1)$, from which we may estimate the analyte mass.

After the n-th measurement (if $n < N$) we make the $(n+1)$-th estimate of the mass:

$$m(n+1) = \widehat{m}(\langle g \rangle)$$

where

$$\langle g \rangle = \frac{1}{n} \sum_{1}^{n} \frac{g(i)y(i)^2}{t} \quad .$$

Then we make the $(n+1)$-th design decision:

$$y(n+1) = \widehat{y}(\widehat{m}(n+1), T)$$

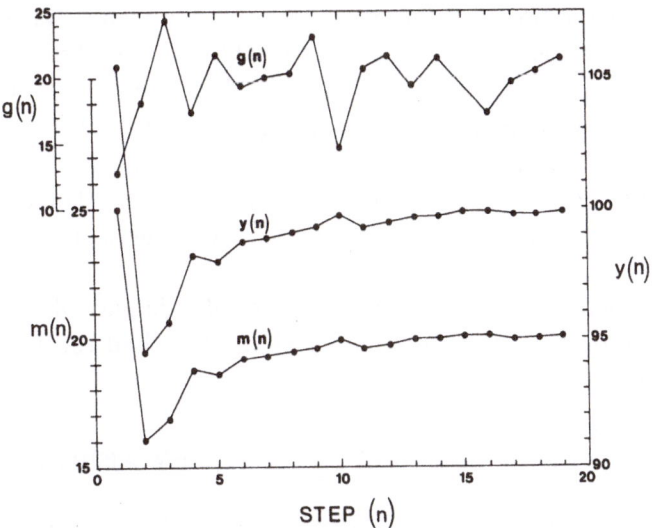

Figure 6.3.1 Adaptive selection of the radial detector position $y(n)$ on the basis of measurements $g(n)$ and estimates $m(n)$ of the analyte mass.

Now the detector is positioned at point $y(n + 1)$ and the next measurement is taken.

It is instructive to consider a numerical example. Let $M = 25$, $V = W = 5$, $s = 10^4$, $K = 10^6$, $T = N = 19$ and $b = 2$. In figure 6.3.1 we show $m(n)$, $y(n)$ and $g(n)$ versus n. The true mass of analyte in this sample is 20, while the initial guess is 25. This initial guess causes the detector to be positioned initially at point 105.4. The response is low, and the detector position is adjusted to 94.5. In the course of the next few measurements, the estimate of the mass steadily improves, and the detector position settles in on the true optimum value of 99.7. Random fluctuations of the individual measurements continue throughout the assay.

At each stage of the assay we have deliberately chosen $y(n+1)$ as though the entire measurement duration T remains. This requires some explanation. One may be tempted to choose $y(n+1)$ on the basis of the remaining time: $T - nt$. This would lead to an unavoidable convergence, toward the end of the assay, of the

detector to its closest allowed position. This would be unsatisfactory since this design is essentially oblivious to the estimated
mass of the analyte. On the other hand, by choosing the detector
position at the $(n+1)$-th step as though T seconds remain we are
ignoring the fact the nt seconds have been expended in "learning"
the sample thoroughly enough to reach the current estimate of the
mass. This will not be a serious deficiency if the learning period
is short.

This dilemma regarding the choice of $y(n)$ is part of the larger
problem of optimizing a dynamic system. We shall return to this
subject in Chapter 7.

6.4 ADAPTIVE ASSAY OF A PULMONARY AEROSOL

The assay of pulmonary plutonium aerosol is one of the most
difficult measurements performed in the routine health safety monitoring of radiation workers [3]. The measurement is often cor-
rupted by random variations of a number of factors. Random
variability of the counting rate is a particularly acute problem because of the low intensity of the L X-ray or gamma radiation which
is measured. In addition, considerable individual variation occurs
in the structure and composition of the thoracic region. Furthermore, the spatial distribution of the aerosol in the respiratory tract
is subject to large variability. Because of the strong absorption of
the measured radiation in the lungs and chest wall, variability of
these factors introduces enormous uncertainties in the calibration
of the measurement. For further discussion see section 3.7.

Many instrumental techniques have been developed to combat
the effects of random variation of the factors mentioned [4]. The
assay-system designer has learned to employ different tools for
different specific situations. Very often, however, a given assay-
system will not be optimal for the full range of variability of the
subjects which the system is intended to measure. As a very simple
example, if a given individual from a group of workers has a very
heterogeneous spatial distribution of the source material, it will
probably be necessary to measure from several positions around

the subject. However, if the total amount of source material is large, each measurement can be of very short duration and the detector can be placed at some distance from the subject. On the other hand, a different individual may contain a small quantity of source material which is nevertheless distributed homogeneously. In this case what is needed is a single measurement of long duration with the detector close to the subject.

With the aid of a micro-computer, one is able to employ the technique of adaptive assay to successfully manage a large range of subject variability. In an adaptive assay system the measurements are interpreted on-line, so as to "learn" the details of the internal structure, composition and source-distribution in the subject being measured. The results of this interpretation are fed back to the assay system to cause computer controlled on-line adaptive modification of the assay-system configuration.

6.4.1 The Model

The pulmonary region will be modelled as a cylinder. The aerosol is distributed in the pulmonary region as a cloud of uncertain shape, whose centroid is eccentrically located in the horizontal midplane at an angle ϕ with respect to an arbitrary line of reference. We shall assume that the detector can be located at any angle θ with respect to the reference line, and that the detector efficiency is independent of angle.

The average response (number of counts) obtained above the background, in a duration t from a detector at angle θ, depends on the total activity A of the aerosol and on its angular orientation ϕ. We shall assume that the functional form of this average response is

$$\langle g|A, \phi; \theta, t \rangle = sAt\big(2 + \cos(\phi - \theta)\big) \quad .$$

Thus the average response increases with the aerosol activity and with the degree of angular alignment between the aerosol cloud and the detector. The parameter s takes a value of 1 ct/nCi-min., and the activity ranges from 0 to about 10 nCi.

We shall assume that the response g has a Poisson distribution with this conditionl mean (abbreviated by $\langle g \rangle$). The conditional probability distribution of g is therefore

$$p(g|A, \phi; \theta, t) = \frac{\langle g \rangle^g e^{-\langle g \rangle}}{g!} \quad , \quad g = 0, 1, 2, \ldots \quad .$$

Furthermore we shall assume that n measurements of equal duration taken from angles $\theta_1, \ldots, \theta_n$ are statistically independent. Thus the joint conditional probability distribution is the product of the individual distributions.

From this formulation of the assay problem several points become clear. First, a single measurement is insufficient for determining the aerosol activity if the angular orientation of the aerosol cloud is unknown. This difficulty has no relation to the statistical uncertainty of the measurement and cannot be overcome by extending the duration of measurement. Rather it arises from the unknown spatial distribution of the aerosol. The second point is that since the activity and angular orientation of the aerosol are independent model parameters, a single measurement is unlikely to be able to resolve both of them. Two or more measurements, from different positions, are going to be needed to unravel this assay problem. The third point concerns the optimal choice of measurement angles. The aim of reducing the statistical uncertainty of the measurement conflicts with the aim of reducing the spatial uncertainty. The statistical uncertainty may be reduced by maximizing the response. This calls for placing the detector opposite the centroid of the aerosol. However, in order to resolve the spatial uncertainty of the aerosol distribution it is necessary to measure from different angles around the subject. This means that not all the measurements will have maximal response.

It is clear from these considerations that we must use rigorous and physically meaningful design and interpretation algorithms in order to achieve an efficient and accurate assay.

6.4.2 The Design Algorithm

The design algorithm by which the measurement angles are selected is based on the Rao-Cramer inequality and its multi-dimensional extension employing the Fisher information matrix (see section 5.3). The logarithm of the joint conditional probability distribution of N measurements, each of duration t, is

$$
\ln p(g_1, \ldots, g_N | A, \phi; \theta_1, \ldots, \theta_N, t)
$$
$$
= \sum_{i=1}^{N} \ln p\left(g_i | A, \phi; \theta_i, t\right)
$$
$$
= \sum_{i=1}^{N} \left(g_i \ln\left(sAt(2 + \cos d_i)\right) - sAt(2 + \cos d_i) - \ln g_i!\right)
$$

where
$$
d_i = \phi - \theta_i \quad , \quad i = 1, 2, \ldots, N \quad .
$$

In order to construct the Fisher information matrix we must evaluate the expectation (with respect to the responses) of the second partial derivatives of $\ln p$. These are readily found to be

$$
q_{11} = -\mathrm{E}\left(\frac{\partial^2 \ln p}{\partial A^2}\right) = \frac{st}{A} \sum_{i=1}^{N}(2 + \cos d_i)
$$

$$
q_{22} = -\mathrm{E}\left(\frac{\partial^2 \ln p}{\partial \phi^2}\right) = sAt \sum_{i=1}^{N} \frac{\sin^2 d_i}{2 + \cos d_i}
$$

$$
q_{12} = q_{21} = -\mathrm{E}\left(\frac{\partial^2 \ln p}{\partial A \partial \phi}\right) = -st \sum_{i=1}^{N} \sin d_i
$$

Suppose for the moment that we know the angular orientation of the aerosol centroid. The Rao-Cramer inequality states that the variance of an unbiased estimate for the aerosol activity can be no less than

$$\mathrm{var}_{rc}(A) = \frac{1}{q_{11}} \quad .$$

The optimal design is the one for which this Rao-Cramer lower bound is as small as possible. This design algorithm (which is based on knowledge of ϕ) directs us to choose

$$d_i = 0 \quad \text{or} \quad \theta_i = \phi \quad , \quad i = 1, 2, \ldots, N \quad .$$

Now let us suppose that the aerosol activity is known. The Rao-Cramer lower bound for the variance of an unbiased estimate of the angular orientation ϕ of the aerosol is

$$\mathrm{var}_{rc}(\phi) = \frac{1}{q_{22}} \quad .$$

This lower bound for the variance of the estimate of ϕ is minimized by choosing

$$\cos d_i = -2 + \sqrt{3} \quad \text{or} \quad \theta \approx \phi \pm 1.84 \text{ radians}$$

for $i = 1, 2, \ldots, N$. This design algorithm (which is based on knowledge of A) directs us to place the detectors at positions which are far from optimal for estimating A (given ϕ).

One way to resolve this conflict is to recognize that we know neither A nor ϕ. In this case, the diagonal elements of the inverse of the Fisher information matrix provide lower bounds for the variance of unbiased estimates of A and ϕ. These lower bounds are

$$\mathrm{var}_f(A) = \frac{q_{22}}{|Q|}$$

$$\mathrm{var}_f(\phi) = \frac{q_{11}}{|Q|}$$

where $|Q|$ is the determinant of Q. The design algorithm should indicate how to position the measurement so as to minimize these variances.

Let us consider the optimal positioning of two measurements. That is, let $N = 2$. Numerical examination of $\mathrm{var}_f(A)$ shows that minimum variance for the estimation of A is obtained when

$$d_1 = 0 \qquad \text{and} \qquad d_2 \neq d_1 \quad .$$

Likewise the minimum variance for the estimate of ϕ is obtained when

$$d_1 = 0 \qquad \text{and} \qquad d_2 \approx 1.95 \text{ radians} \quad .$$

We can now formulate the design algorithm for selecting measurement angles for a pair of measurements. Given an estimate $\widehat{\phi}$ of the angular orientation of the aerosol centroid, we shall position the measurements at

$$\theta(1) = \widehat{\phi} \qquad , \qquad \theta(2) = \widehat{\phi} + 1.95 \text{ radians} \quad .$$

This is the best choice (in a minimum variance sense) of the measurement angles given uncertainty in both A and ϕ.

6.4.3 The Interpretation Algorithm

The task of the interpretation algorithm is to indicate best estimates $\widehat{A}$ and $\widehat{\phi}$ on the basis of a set of measurements taken from known measurement angles. We shall employ the technique of maximum likelihood estimation (see section 4.4.2). That is, given N measurements, the estimates $\widehat{A}$ and $\widehat{\phi}$ are those values which maximize the joint conditional probability distribution of these measurements. The search for the maximizing values of A and ϕ can be efficiently performed numerically. In fact it is convenient to maximize the following function:

$$H(A, \phi) \equiv \ln p + \sum_{i=1}^{N} \ln g_i!$$

$$= \sum_{i=1}^{N} g_i \ln \left(sAt(2 + \cos d_i) \right) - sAt \sum_{i=1}^{N} (2 + \cos d_i) \quad .$$

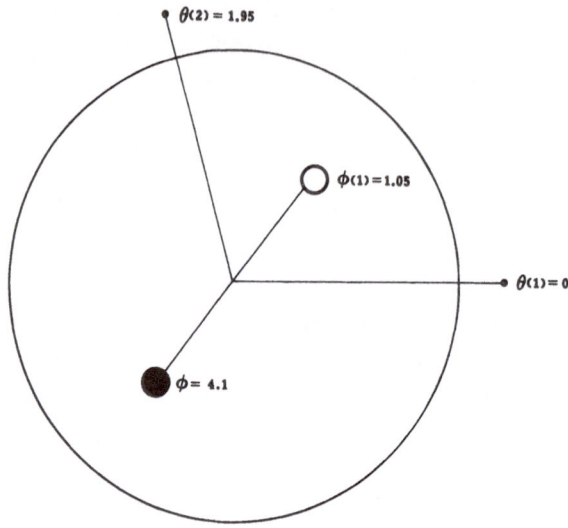

Figure 6.4.1 Assay of a Pu aerosol centered at 4.1 radians. The small solid circle shows the true position of the aerosol, while the small open circle is at the position estimated on the basis of the first two measurements.

6.4.4 Implementation

Let us consider a simulated implementation of this adaptive assay. We shall measure a plutonium aerosol whose activity is 3 nCi. A single detector is to be employed to perform six successive measurements, each of 10 minutes duration.

The design algorithm indicates that if the angular orientation of the aerosol centroid is ϕ radians, then the best angles for a pair of measurements are ϕ and $\phi + 1.95$. However no knowledge is available about the angular orientation of the aerosol before the first measurement. Thus the first two measurements are performed at angles $\theta(1)$ and $\theta(2)$ where

$$\theta(1) = 0 \quad \text{and} \quad \theta(2) = 1.95 \text{ radians} \quad .$$

When this first pair of measurements is complete, the values of A and ϕ are estimated. These estimates are denoted $\widehat{A}(1)$ and $\widehat{\phi}(1)$.

The next pair of measurements are positioned optimally on the basis of this estimate of ϕ. That is

$$\theta(3) = \widehat{\phi}(1) \qquad \text{and} \qquad \theta(4) = \widehat{\phi}(1) + 1.95 \text{ radians} \quad .$$

Upon completion of the second pair of measurements, new estimates $\widehat{A}(2)$ and $\widehat{\phi}(2)$ are obtained from all four measurements. The third and final pair of measurements are made at angles

$$\theta(5) = \widehat{\phi}(2) \qquad \text{and} \qquad \theta(6) = \widehat{\phi}(2) + 1.95 \text{ radians} \quad .$$

The final set of estimates, $\widehat{A}(3)$ and $\widehat{\phi}(3)$, are based on all six measurements. The following table shows the results of numerical simulations. Each column represents the results of a complete assay for an aerosol whose true angular orientation is indicated at the top of the column. The most interesting run is that for $\phi = 4.1$ radians. In this case the aerosol centroid is located quite near the line bisecting the angle between the first two measurements. The situation after the first pair of measurements is shown in figure 6.4.1. After the first pair of measurements, the activity A is underestimated by more than a factor of 2, and the aerosol centroid is put on the wrong side of the subject ($\widehat{\phi}(1)$ is wrong by π radians). This error is due primarily to the inherent spatial ambiguity of the assay problem. When the detectors are repositioned, the estimates improve dramatically. In all of the five assays shown the final estimate of A is off by no more than 10%, and the error in the final estimate of ϕ is no greater than 15%.

$\phi(\text{true}) =$	0	1.57	3.0	4.1	5.0
$g(1)$	100.00	50.00	22.00	35.00	62.00
$\theta(1)$	0.00	0.00	0.00	0.00	0.00
$g(2)$	40.00	76.00	64.00	37.00	36.00
$\theta(2)$	1.95	1.95	1.95	1.95	1.95
$\widehat{\phi}(1)$	5.24	2.84	2.84	1.05	4.64
$\widehat{A}(1)$	4.00	2.20	2.30	1.40	3.20
$g(3)$	76.00	60.00	97.00	24.00	76.00
$\theta(3)$	5.24	2.84	2.84	1.05	4.64
$g(4)$	88.00	30.00	58.00	85.00	68.00
$\theta(4)$	0.87	4.79	4.79	3.00	0.31
$\widehat{\phi}(2)$	6.13	1.65	3.14	3.89	5.39
$\widehat{A}(2)$	3.20	2.60	2.90	2.80	2.80
$g(5)$	80.00	84.00	76.00	80.00	78.00
$\theta(5)$	6.10	1.65	3.14	3.89	5.39
$g(6)$	44.00	56.00	50.00	50.00	48.00
$\theta(6)$	1.80	3.60	5.09	5.84	1.05
$\widehat{\phi}(3)$	6.13	1.80	3.29	4.04	5.54
$\widehat{A}(3)$	3.00	2.70	2.80	2.80	2.70

6.5 ADAPTIVE ASSAY OF A URANIUM DEPOSIT

The adaptive approach to assay is essential in those situations where lack of prior knowledge about the sample precludes pre-measurement design optimization. The assay of a heterogeneous subterranean uranium deposit is a clear case in point. The need for an adaptive assay of a uranium deposit arises from several factors. The geomorphology of such deposits is quite variable, and is characterized by the overall size and shape of the deposit, by the size, mass and spatial distribution of the uraniferous inclusions, by the average radiation-absorption characteristics of the non-uraniferous matrix as well as the degree of heterogeneity of the matrix. Spatial heterogeneity of the deposit typically introduces large spatial uncertainty to the determination of the average uranium concentration. Proper assay design is therefore quite important, and what is particularly problematic is that the best design is sensitive to details of the geomorphology.

Statistical uncertainty is also a prevalent problem in uranium assay. The source intensity is low and the count rate of background radiation may be many times that of the uranium.

A further incentive for employing an adaptive technique is that the range of realizable design options is severely limited, so it is important to effectively utilize the available instrumental potential. The duration of measurement is usually limited. The sample is typically of effectively infinite extent in at least two dimensions, and measurements can be made only along the length of a single bore-hole. Thus the powerful assortment of detector-sample configurations employed in above-ground assay of finite samples is almost entirely beyond the bounds of practical implementation. The use of high-resolution solid state detectors is difficult (though not impossible) and standard measurements are based on more rugged NaI scintillation detectors with poor energy resolution.

6.5.1 The Model

We shall develop an adaptive assay for the thin heterogeneous uraniferous layer described in section 5.2.2. Two model parameters characterize the geomorphology of the layer: the average uranium concentration w (g/cm^3) and the mass M (grams) of the uniform uranium nodules. The entire deposit contains several parallel layers, separated one from the other by thick layers of non-uraniferous material. From independent geological evidence we are able to assume that the same values of M and w characterize all the layers in the deposit. Furthermore, this value of M is likely to range from 0.01 to 1.0 g. The thickness of each layer is 30 cm. The aim of the assay is to determine the average uranium concentration in the deposit.

These two model parameters, M and w, determine all the available statistical information about the uranium deposit, as embodied in the mean and variance of the detector response g. For a single detector located at a distance z from the center of the deposit, the mean and variance of the number of counts g obtained in a duration t are expressed in eqs. (5.2.5.) – (5.2.8). These relations may be summarized as

$$\langle g|M, w; z \rangle = wt f_1(z) \qquad\qquad (6.5.1)$$
$$\text{var}(g|M, w; z) = Mwt^2 f_2(z) + (1 + b)wt f_1(z) \qquad (6.5.2)$$

where f_1 and f_2 are known functions of the detector position and b is the ratio of the background to the uranium count rate. The first and second terms on the righthand side of eq.(2) are the spatial and statistical variance of g, respectively. The functions $f_i(z)$ are shown in figure 6.5.1 for a bore-hole radius of 3 cm, a linear absorption coefficient of 0.13 cm^{-1}, and an overall detector efficiency of 1.6×10^4, as in section 5.2.2.

6.5.2 The Design Algorithm

The model of the uranium deposit contains rather limited information, as expressed by the mean and variance of the response rather than by a complete probability distribution. Consequently the design algorithm will be based on the relative error of g. The aim of the design algorithm is to indicate the position, z, from which to measure each layer of the deposit, in the light of estimates of the model parameters M and w. The detector is to be positioned at that point for which the relative error is a minimum.

Examination of eqs. (1) and (2) shows that the choice of the optimal position is independent of the value of w, since the relative error is just inversely proportional to $w^{1/2}$. However, the choice of the optimal detector position is quite sensitive to the value of M, the uranium nodule size. Figure 6.5.2 shows the relative error versus the detector position (relative to the center of the layer) for

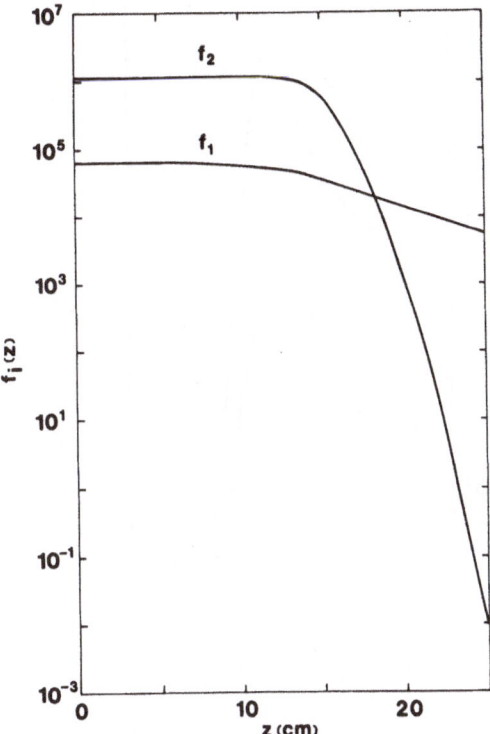

Figure 6.5.1 Functions $f_i(z)$, for evaluation of the conditional mean and variance of the detector response.

various values of M. The uranium concentration is $3 \times 10^{-4} \mathrm{g/cm}^3$ and the measurement duration is 40 seconds. When M is large, the number density of nodules N nodules/cm^3 is low, since w, M and N are related by

$$w = NM$$

As a consequence, a large value of M implies that the spatial distribution of uranium is very heterogeneous. Thus the spatial uncertainty is greater than the statistical uncertainty, and a global minimum in the relative error occurs outside the uranium layer. Indeed when $M = 1$ g, the relative error at the center of the layer is 0.93, and at 5 cm outside the layer ($z = 20$ cm) the relative error

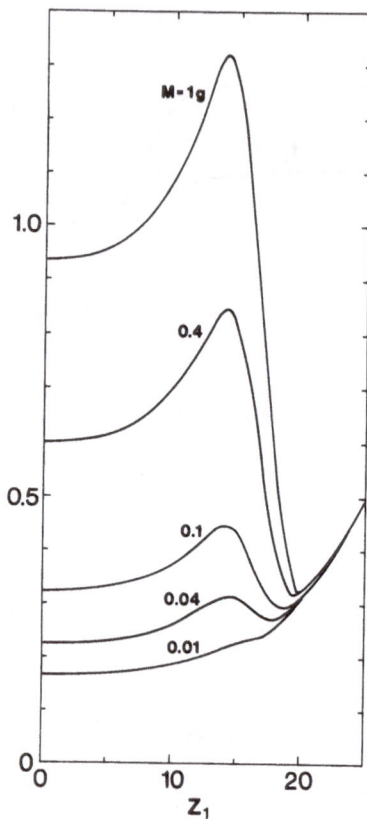

Figure 6.5.2 Relative error of the response versus the detector position
with respect to the center of the layer, for various nodule masses.

is 0.32. At the other extreme, when the nodule mass is small, the
number density is large and the uranium is fairly homogeneous.
In this case the statistical uncertainty is dominant, and may be
minimized by measuring at the center of the layer. For $M = 0.01$
g, the relative error at $z = 0$ is 0.17, and at $z = 20$ cm it takes
a value of 0.31. Thus for small particles the optimal detector
position is at the center of the layer.

Figure 6.5.3 summarizes the design algorithm by showing the
ratio of the relative error at $z = 0$ to the relative error at the local

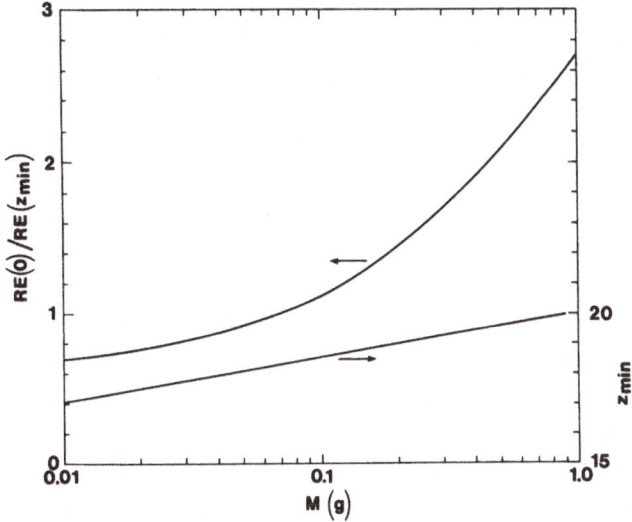

Figure 6.5.3 The upper curve shows the dependence on M of the ratio of the relative error at the center of the layer to the relative error at the local minimum outside the layer. The lower curve shows the position of the local minimum.

minimum outside the layer. When this ratio equals unity we are indifferent as to where the detector is located. When this ratio exceeds unity the detector should be positioned outside the layer at the point of the local minimum, z_{min}, which is a function of M. When this ratio is less than one, the detector is to be positioned at the center.

The design decision will be based on an estimate of M which is likely to be rather crude. Examination of figure 6.5.2 shows that it is preferable to err in the design decision when M is small rather than when M is large. That is, to measure from 20 cm when $M = 0.01$ introduces a less serious error than to measure from the center when $M = 1$. The ratio of relative errors equals unity at $M = 0.07$ g. In the interest of caution we shall adopt a lower threshhold for measuring from above the layer. Specifically, the design algorithm is that the measurement will be performed outside the layer if $M > 0.05$ g.

The position outside the layer of the local minimum in the relative error is also shown in figure 6.5.3. This dependence of z_{min} on M can be well approximated by

$$z_{min} = 20 + 0.65 \ln M$$

6.5.3 The Interpretation Algorithm

Only 60 seconds of measurement time are available for assaying each layer of the deposit. It has been decided that 20 seconds are to be employed on each layer in order to estimate M, and the remaining 40 seconds will be used to assay w. From eqs. (1) and (2) it is evident that in order to estimate M it is necessary to estimate the spatial variance of g. This requires a number of spatially independent measurements with low statistical uncertainty. Spatial independence of two measurements requires that they be separated sufficiently from one another so that their effective fields of view are not overlapping. On the other hand, to minimize the statistical uncertainty of the measurements it is necessary to measure from well within the boundaries of the layer. Since the mean free path of the radiation is about 7.7 cm and the total thickness of each layer is 30 cm, it is impossible to perform more than two spatially independent measurements inside the layer. Two measurements are hardly adequate to determine the spatial variance of the response. However, we may exploit the fact that the nodule mass is the same in all the layers of the deposit. Thus the proposed measurement scheme is to use a two-pass approach. On the first pass the detector is lowered through the deposit and performs two 10-second measurements in each layer, one at $+10$ cm and one at -10 cm with respect to the center of the layer. If K is the number of layers, the resulting $2K$ measurements will be used to estimate M and to determine the optimal detector position for the subsequent measurements. In the second pass the detector is drawn out through the layers and performs 40 seconds of measurement in each layer. If the optimal detector position, as determined by the first-pass measurements, is $z = 0$ cm, then a single 40-second measurement is made in each layer. If the optimal detector position is outside the layer, then two 20-second measurements are made, one below and one above the layer. The resulting measurements are used to estimate w.

The actual estimates of M and w are straigtforward applications of eqs. (1) and (2). Suppose that we have n spatially and temporally independent measurements of a deposit. Each measurement is of duration t seconds and is from position z relative to the center of the layer in which the measurement was made. Let $\langle g \rangle$ and $\text{var}(g)$ be the experimental average and variance of the n measurements. The estimates $\widehat{w}$ and $\widehat{M}$ are obtained from the relations

$$\langle g \rangle \equiv \frac{1}{n} \sum_{i=1}^{n} g_i = \widehat{w} t f_1(z)$$

$$\text{var}(g) \equiv \frac{1}{n} \sum_{i=1}^{n} (g_i - \langle g \rangle)^2 = \widehat{M} \widehat{w} t^2 f_2(z) + (1 + b) \widehat{w} t f_1(z)$$

In the first pass, the number of measurements will be $2K$, twice the number of layers, t will be 10 seconds and z will be 10 cm. In the second pass, the number of measurements will be K or $2K$ and t will be 20 or 40 seconds, depending on whether z is zero or $z_{\min}$.

6.5.4 Implementation

Let us consider the simulated implementation of this adaptive assay scheme. The deposits to be measured have a true uranium concentration of 3×10^{-4} g/cm^3, and are comprised of 5 separate 30 cm layers. Each line of tables 6.5.1 and 6.5.2 shows the results of the two-pass assay of such deposits. Table 6.5.1 shows the assay of 10 deposits, each comprised of five layers, and each with a uranium nodule mass of $M = 0.5$ g. Table 6.5.2 shows the assay

of 10 multi-layer deposits with $M = 0.01$ g. The first two columns
of each table show the estimated values of w and M after the first
pass through the five layers. The third column shows the detector
position chosen for the second-pass measurements. The fourth and
fifth columns show the estimates of w and M on the basis of the
second pass.

Table 6.5.1

Adaptive Assay of Uranium Deposits
$$w = 3 \times 10^{-4} \text{g/cm}^3$$
$$M = 0.5 \text{g}$$

$\widehat{w}(1)$ $(\times 10^4)$	$\widehat{M}(1)$	$z(2)$	$\widehat{w}(2)$ $(\times 10^4)$	$\widehat{M}(2)$
1.96	0.25	19.1	3.26	−0.59
4.88	0.30	19.2	3.34	0.0087
3.58	0.18	18.9	3.47	−1.1
4.36	0.25	19.1	2.97	1.9
3.35	0.17	18.8	3.19	1.1
4.53	0.27	19.2	2.87	−0.22
2.96	0.20	19.0	2.88	−0.34
3.30	0.29	19.2	3.05	−0.021
2.76	0.38	19.4	3.43	2.0
3.74	0.37	19.4	3.06	−0.72

Several points are immediately evident. The first-pass estimates of M are crude. However, they are sufficiently accurate
to allow the correct design decision (between measuring at $z = 0$
or outside the layer) to be made in nearly all cases. The first-
pass estimates of w are quite variable, while the second-pass estimates are more stable. The second-pass estimates of M in the
large-nodule deposits (table 6.5.1) are so far off as to be nearly
meaningless. This apparently arises from the fact that, since the
second-pass measurements are made from outside the layer, the

Table 6.5.2

Adaptive Assay of Uranium Deposits
$$w = 3 \times 10^{-4} \text{g/cm}^3$$
$$M = 0.01 \text{ g}$$

$\widehat{w}(1)$ $(\times 10^4)$	$\widehat{M}(1)$	$z(2)$	$\widehat{w}(2)$ $(\times 10^4)$	$\widehat{M}(2)$
3.00	−0.0058	0.0	3.33	−0.0089
2.70	0.0046	0.0	2.83	0.021
2.56	0.023	0.0	3.09	0.0056
3.35	0.073	18.3	3.15	−0.028
3.26	0.11	18.6	3.15	−0.053
3.48	−0.017	0.0	3.20	0.012
3.33	−0.0068	0.0	3.08	0.0075
3.31	0.051	18.1	2.69	0.27
2.29	0.0017	0.0	2.87	−0.0088
2.94	0.0013	0.0	3.05	−0.0014

statistical variance of g is much greater than the spatial variance, so the estimate of M is quite unreliable. Table 6.5.3 summarizes the overall performance of the adaptive assay in each of the two sets of deposits ($M = 0.01$ and $M = 0.5$ g). The averages and the standard deviations of the estimates of w are shown for both the first and second pass. Also shown is the average squared-deviation of the estimated w from the true value of w. These data indicate that the second-pass measurements are consistently and appreciably better than the first-pass measurements.

In order to decide that the adaptive assay is really preferable, we must compare it with some reasonable non-adaptive assay schemes. The most likely non-adaptive approach is to make all measurements from outside the layer. This will be nearly optimal when $M > 0.05$ g, and sub-optimal otherwise. However, recalling figure 6.5.2, we note that the consequences of non-optimality when

Table 6.5.3

Summary of Adaptive Assay of Uranium Deposits

	$M = 0.01$ g	$M = 0.5$ g
$\langle w(1) \rangle$	3.022	3.54
$\sigma_{\widehat{w}(1)}$	0.38	0.84
$\langle (\widehat{w}(1) - w_{tr})^2 \rangle$	0.14	1.01
$\langle w(2) \rangle$	3.04	3.15
$\sigma_{\widehat{w}(2)}$	0.18	0.21
$\langle (\widehat{w}(2) - w_{tr})^2 \rangle$	0.035	0.078

M is small are less severe (though not negligible) than when M is large. An alternative non-adaptive scheme is to make all the measurements near the center of the layer. This will be optimal for $M < 0.07$, but far from optimal otherwise.

In table 6.5.4 we show the simulated results of assaying 10 multi-layer deposits, where the true values of w and M are 3×10^{-4}g/cm^3 and 0.01 g respectively. Each deposit is comprised of 5 layers as before, and each layer is assayed for 30 seconds at $z = 19$ cm and for 30 seconds at $z = -19$ cm. The average and standard deviation of the estimates of w, and mean squared-deviation from the true value of w are also shown. For the adaptive assay of deposits with $M = 0.01$ g the standard deviation of the estimate of w is 0.18, while in this non-adaptive assay the standard deviation is larger by 50%, being 0.27. The adaptive assay is more reliable than this non-adaptive assay.

The second column of table 6.5.4 shows the simulated results of assaying 10 multi-layer deposits for which the true value of w is as before and $M = 0.5$ g. In this case the second-pass estimate from the adaptive assay is far more reliable than this non-adaptive approach. The standard deviations of the adaptive and non-adaptive estimates of w are 0.21 and 0.82 respectively.

Table 6.5.4

Non-Adaptive Assay of Uranium Deposits

$M = 0.01$ g	$M = 0.5$ g
$\widehat{w}$	$\widehat{w}$
2.59	2.09
3.19	2.21
3.29	3.12
3.35	3.68
2.83	3.63
3.41	4.78
3.16	4.30
3.06	3.01
2.68	3.04
3.07	2.57

$$\langle \widehat{w} \rangle = 3.06 \qquad \langle \widehat{w} \rangle = 3.24$$
$$\sigma_{\widehat{w}} = 0.27 \qquad \sigma_{\widehat{w}} = 0.82$$
$$\langle (\widehat{w} - w_{tr})^2 \rangle = 0.074 \qquad \langle (\widehat{w} - w_{tr}) \rangle = 0.74$$

The conclusion is that, if one expects a wide range of uranium nodule sizes, the adaptive approach generally yields a superior estimate of the average uranium concentration.

The geomorphological model employed in this example can be extended in many ways, some of which were mentioned at the start of the section. Of particular importance for actual implementation of the adaptive assay discussed here is determination of the true stratigraphy of the deposit. In practice it is necessary to adaptively find the boundary of each layer. In view of the complex structure of such deposits, the adaptive approach to assay is imperative. The rewards, in terms of assay accuracy, can be quite significant.

NOTES

[1] See refs. [1.4] and [2.1] of Chapter 5. Also see
A. Wald, *Sequential Analysis*, John Wiley, 1947.

[2] J. O. Berger, *Statistical Decision Theory*, Springer Verlag, 1980.

[3] Y. Ben-Haim and A. Kushelevsky, Assay of Pu-Aerosol in Human Lungs — Instrument Design and Data-Interpretation, IAEA-WHO Internatl. Symp. on Assessment of Radioactive Contamination in Man, paper IAEA-SM-276/3, Paris, Nov. 1984.

[4] Numerous papers in the Symposium of ref. [3] discuss various instrumental methods for assay of internal radioactive contamination. For examples, see
1. R. C. Lane *et al*, The Use of Six-Element Arrays of Hyperpure Ge Detectors in Monitoring for Internal Actinide Contamination. Paper IAEA-SM-276/24.
2. T. Rahola *et al*, The Advantage of Using A Semiconductor Detector for Determination of Internal Radioactive Contamination. Paper IAEA-SM-276/10.
3. C. Pomroy and H. Malm, Hyperpure Ge Detectors for *In Vivo* Measurements of U and Th. Paper IAEA-SM-276/49.
4. D. Hernandez, M. Righetti and J. Chagaray, Measurement of ^{239}Pu and ^{241}Am with CsI(Tl) and I_2Hg Detectors. Paper IAEA-SM-276/64.
An comparison of detector types may be found in:
5. S. Mizushita, Relative Sensitivity of Hyperpure Ge Detector Array and Phoswich Detectors in Assessment of Pu and Am in Lungs, *J. Nucl. Sci. Tech.*, 21:775-85 (1984).

CHAPTER 7

SOME DIRECTIONS FOR RESEARCH

7.1 OVERVIEW

Two topics have occupied our attention throughout this book: assay-system design and data-interpretation. We have developed a set of analytical tools to assist the designer in making rational decisions, and we have studied a range of techniques for extracting information from measurements which are distorted by random processes of various sorts. The tasks of design and interpretation have been synthesized by the concept of adaptive assay, with the result that the power of each individual group of techniques is greatly augmented.

Three concepts of much deeper origin have implicitly guided our way: optimality, realizability and generality. We have sought to define a "best" design and a "best" estimate. In addition, the criteria of optimality have been chosen so that the details of the optimal solution can be realized with limited resources such as time and information. Attention to the problem of realizability is particularly important in the adaptive synthesis of design and interpretation.

Our formulations have been given a deliberate generality. However, generality is not a goal in itself, as are optimality and realizability. Rather, generality is a problem-solving technique. We are concerned with solving practical problems whose complexity, when considered in detail, is often overwhelming. Complicated problems, when formulated in a general fashion, sometimes engender quite tractable solutions. The technique of design-analysis based on the convexity theorem is a clear example.

It is commonly recognized that by casting a problem in an abstract form, existing solutions or techniques from apparently

297

unrelated fields may become relevant. The converse is also true: the general solution of one problem may find applications far afield. One of the aims of this final Chapter is to indicate some directions in which this converse is manifested by the concepts which we have developed.

Challenging questions also remain in the area of material assay. A selection of such problems are included in this Chapter.

7.2 NON-LINEARITY IN MASS

Two fundamental linearity properties of detector response were introduced in section 2.2. These are linearity in time and linearity in mass. Linearity in time states that the number of counts to be expected in an interval t is precisely t times the expected number of counts in a unit interval of time. For a large class of assay problems this is quite a reasonable assumption. The property of linearity in mass asserts that the response to a spatial distribution $r(x)$, for an assay system whose point-source response functions is $f(x)$, is

$$c(r) = \int r(x) f(x) \, dx$$

where the integral is over the volume of the sample.

Important classes of assay systems display linearity in mass, as has been shown by numerous examples throughout the book. However not all assay problems are so well behaved. For example, if the radiation transport characteristics of the analyte are substantially different from those of the matrix material, and if the analyte is present in sufficient concentration to alter the overall transport properties of the sample, then the response to the spatial distribution $r(x)$ will not be a simple linear superposition.

The most serious problem which arises from non-linearity in mass is that the complete response set is no longer the convex hull of the point-source response set. As a consequence, evaluation of the relative resolution becomes quite difficult. The simple and efficient min-max algorithm for evaluating the expansion of the complete response set may no longer be applicable.

We have encountered a related problem in assay problems for which the class of complete response sets does not display the property of proportionality. Further investigation of non-proportional response sets may shed light on the problem of non-linearity in mass. In more general terms, non-linearity in mass invites the application of techniques from the growing field of non-linear mathematics [1].

7.3 NON-CONVEX RESPONSE SETS

Convexity of the complete response set is a cornerstone of our analysis. If the complete response set is non-convex, the evaluation of the relative resolution may become quite complex. It is thus important to develop methods to handle such situations.

However, non-convexity does not necessarily invalidate the utility of the min-max algorithm for calculating the relative resolution. For instance, if the complete response set has the form of two tangent circles whose centers lie on a ray from the origin, it will be easy to find the relative resolution. This complete response set is not convex, but it is locally convex at certain (crucial) parts of its boundary. It would be important to establish criteria for sufficient convexity and techniques for evaluating the relative resolution for such assay systems.

7.4 ASYMPTOTIC DESIGNS

The relative resolution has proven to be a powerful design tool in the analysis of multi-detector assay systems. A typical question which arises in the design of such systems is to determine the utility of the marginal (n-th) detector. The relative resolution is a well-suited measure of performance for addressing this question. We have examined a number of applications where the relative resolution is evaluated for an ever increasing number of detectors. As the detector multiplicity rises, the relative resolution converges to a limiting value. This asymptotic value of the relative resolution, which may exceed unity, defines the best resolution attainable from the design-concept which underlies the sequence of designs which have been evaluated.

The evaluation of the relative resolution for multi-detector systems can be efficiently computerized. Nevertheless the computation involved in calculating the asymptotic resolution can sometimes be time consuming. When comparing different multi-detector design concepts it may be desirable to know the asymptotic value of the resolution which characterizes each concept. The computational effort required for comparison of several distinct concepts may be large indeed.

In light of these considerations, it would be useful to have a technique for direct evaluation of the asymptotic relative resolution. By direct evaluation we mean that the asymptotic value is found by a single calculation, not "empirically" as the numerical limit of a sequence of multi-detector calculations.

7.5 MALFUNCTION ISOLATION

The approach to optimal design of assay systems based on the concept of relative resolution has a promising application to the problem of isolating a malfunction in a complex dynamical system. More precisely, there is a formal parallelism between the design of an assay system for spatially random material and the design of a filter for malfunction isolation.

In order to introduce this parallelism, we begin by formulating the malfuncton isolation problem for linear systems. Let x be the state vector, and let the normal dynamics be represented in matrix form by a set of linear differential equations as

$$\frac{dx}{dt} = Ax \quad .$$

Control and noise vectors are ignored for simplicity. The system is measured, and the measurement vector is

$$y = Hx \quad .$$

If a malfunction occurs in the dynamics of the system, then the matrix A changes. The dynamic equations become

$$\frac{dx}{dt} = (A + F)x \quad .$$

Similarly, if sensor failure occurs, the matrix H changes. Non-homogeneous "bias" terms may also be introduced by malfunction.

The aim of a malfunction isolation filter is to determine which elements of A (or at least which rows of A) have changed, and by how much. This may be viewed as an assay problem. The "sample container" is the failure matrix F. That is, if F is a matrix of dimension N, then the spatial domain of the sample is the collection of N^2 positions in the failure matrix. The analyte "material" is the value of the elements of F. The malfunction isolation filter must determine how much "material" is concentrated in which elements of F. For a simple failure, only a single element of F is non-zero, so zero- and first-moment assays completely determine the failure. For more complex failures, higher order moments are needed for complete characterization of the failure.

We are not interested in the isolation filter *per se*. What we are after is a technique for objectively comparing the resolution capability of alternative filters. These filters need not be based on a "moments" method at all. However, it should be possible to formulate an arbitrary isolation filter as a moment assay. If this is so, then the filter can be objectively and quantitatively compared with other filters by means of the measure of performance. Furthermore, the measure of performance is a parameter which has a physically meaningful best value. Consequently we are able to define an optimal malfunction isolation filter. This is analogous to the classical estimation problem for linear systems with Gaussian noise, for which the Kalman filter is a best estimate in the maximum likelihood sense [2].

A simple example is in order. A malfunction isolation filter is usually based on a hypothesis of the form of the malfunction [3]: what elements of the matrix A change and how they change in time. A sophisticated algorithm may encompass a bank of parallel filters, each based on a different failure hypothesis. The set of allowed or realizable malfunctions is undoubtedly larger than the set of hypothesized malfunctions in the bank of filters. The measure of performance is able to evaluate how well a given bank of filters will perform when operating in the environment of a given set of allowed possible malfunctions. (Just as the measure of performance of a material assay system is able to evaluate how well a given design will perform in the environment of a given set of allowed spatial distributions of the analyte). Furthermore, the measure of performance allows comparisons of different banks of filters, and thus enables quantitative and physically meaningful answers to questions such as: How does the malfunction-isolation capability obtained from a filter based on one particular failure hypothesis compare with the capability derived from a different hypothesis, for a given range of possible malfunctions? How much better is a bank of 10 filters than a bank of 2, 100 than 10, etc? In general, what is the utility of the marginal (n-th) filter?

The relative resolution seems to provide a concise, comprehensive and computable criterion of optimality for malfunction isolation filters. Many open questions remain concerning its implementation as a design tool.

7.6 ADAPTIVE ASSAY — ADVANCED CONCEPTS

7.6.1 Global Optimization

The formal structure of the adaptive assay scheme is to be found in the theories of adaptive control [4] and dynamic optimization [5]. As such, very powerful analytical tools can be employed in the formulation and implementation of the design-optimization algorithm. In this section we shall describe a general concept of adaptive assay, whose implementation can be facilitated by existing techniques of optimization. It is beyond the scope of this book however to discuss these techniques in any depth.

The problem which one seeks to solve is the overall optimization of the assay. That is, one would like to exploit the available instrumental potential so as to achieve the best possible assessment of the assayed parameters. The difficulty is that knowledge of the sample changes during the assay, and as a consequence the design is modified on-line. An assay configuration which looked optimal early in the assay may seem far from optimal later on. Design decisions made at any stage during the assay should, ideally, account for the sub-optimality of previous designs. In some situations one might even conceive of a design algorithm which in some way estimates the sub-optimality of the current best design. The most comprehensive design algorithm evaluates previous designs in the light of what has been "learned" about the sample, and "looks forward" in an attempt to estimate how much more learning will take place. Such an algorithm is performing "global" optimization of the assay. This is to be contrasted with strictly "local" optimization, in which each step of the adaptive assay is treated as a self-contained independent assay, albeit relying on model-parameter estimates based on previous measurements.

A simple example should clarify the ideas of global and local optimization, and for this we will return to the adaptive assay described in section 6.3. It will be recalled that the assayed material is arbitrarily distributed in the interval $[0, 1]$ on the real line, and that the detector may be located at any point in the interval $[2, \infty)$. The design algorithm is based on the relative mass resolution, which is a function of the detector position y, the total mass m of the analyte and the duration t of measurement. Explicitly, the relative mass resolution with statistical uncertainty was found to be

$$Z(y, m, t) = e^{\mu} \left(\frac{y}{y - 1} + \frac{by}{\sqrt{Kmt}} \right)^2 . \qquad (7.6.1)$$

The optimal detector position is that which minimizes Z, and is found to depend on the estimated analyte mass and on the duration of measurement as

$$\hat{y}(\hat{m}, t) = 1 + \frac{(K\hat{m}t)^{1/4}}{\sqrt{b}} . \qquad (7.6.2)$$

The difficulty of global optimization in this assay arises in choosing the value of t by which the detector position is selected. If the total duration of the assay is T, then the detector should ideally be positioned at $\hat{y}(m, T)$ for the duration of the measurement. However this supposes prior knowledge of the very parameter — m — which is being assayed. In adopting the adaptive approach the total measurement duration is divided into N subintervals of length $\tau = T/N$. After each subinterval the mass is estimated and the detector is repositioned for an additional measurement of duration τ. A strictly local optimization would position the detector for each interval according to the best estimate of m and the length τ of the subinterval. This clearly will be unsatisfactory in a situation in which the estimate of m is stable throughout the assay and fairly good from the start. In this case we would want the detector to be positioned at $\hat{y}(\hat{m}, T)$.

On the other hand, suppose the early estimates of m are poor and variable and that a fair fraction of the total measurement duration is expended in achieving a reasonably stable estimate. Once a somewhat stable estimate of m has been attained, it would be unrealistic to position the detector as though the entire T seconds remain. This would cause the detector to be positioned much too far from the sample for the actual remaining measurement duration, with the consequence that the optimal balance between statistical and spatial uncertainty would not be achieved.

In an attempt to avoid the suboptimality of the above alternatives, one might consider positioning the detector on the basis of the remaining duration; that is, at $\hat{y}(\hat{m}, T - nt)$ after n subintervals. The problem here is that $T - nt$ gets small as n approaches N. Thus the detector is irrevocably forced to the closest position $(y = 2)$ at the end of the assay. This algorithm ignores the estimate of m toward the end of the measurement when $\hat{m}$ is most accurate.

The design algorithm expressed in eqs.(1) and (2) can be implemented in various alternative ways, which achieve a greater or lesser degree of global optimization. However since eq.(1) is the relative mass resolution of only a single measurement, its potential as a tool for global optimization may be limited. Instead of eq.(1) one may base the design algorithm on the relative resolution of the full N-measurement adaptive assay. For example, let $Z(y_1, \ldots, y_N, m, t)$ be the relative mass resolution for N measurements, each of duration t, for analyte mass m. The n-th detector position, for $n \leq N$, can be chosen by minimizing Z with $y_1, \ldots, y_{n-1}$ fixed at the previous detector positions.

From these brief remarks it is clear that development of a global design algorithm is an essential as well as challenging aspect of adaptive assay design.

7.6.2 Sequential Analysis of Adaptive Assay

The duration of measurement can be viewed as a design parameter which can be determined adaptively by the technique of sequential analysis, as discussed in section 6.2. However it is not necessary to limit sequential analysis to adaptive assays in which only the duration is adjusted on-line. A likelihood ratio test can be used to terminate an adaptive assay in which other design parameters are modified as well. In other words, an adaptive assay, such as those illustrated in sections 6.3 through 6.5, can be embedded in a sequential analysis. However, the formulation of the sequential analysis is more complex than that presented in section 6.2 because the sequential measurements are not likely to be independent and identically distributed.

NOTES

[1] 1. T. L. Saaty and J. Bram, *Nonlinear Mathematics*, McGraw-Hill, 1964 and Dover, 1981.
2. T. L. Saaty, *Modern Nonlinear Equations*, McGraw-Hill, 1967, and Dover, 1981.
3. R. Bellman, *Methods of Nonlinear Analysis*, Academic Press, Vol. I 1970, Vol. II 1973.

[2] B. D. O. Anderson and J. B. Moore, *Optimal Filtering*, Pren-

tice-Hall, 1979.

[3] 1. A. S. Willsky, A Survey of Design Methods for Failure
Detection in Dynamic Systems, *Automatica*, 12: 601-11 (1976).
2. L. F. Pau, *Failure Diagnostics and Performance Monitoring*,
Marcel Dekker Pub., 1981.
3. Y. Ben-Haim, Malfunction Isolation in Linear Stochastic Systems: Applications to Nuclear Power Plants, *Nucl. Sci. Eng.*, 85:
155-66 (1983).

[4] 1. R. Bellman, *Adaptive Control Processes: A Guided Tour*,
Princeton University Press, 1961.
2. R. Bellman, *Introduction to the Mathematical Theory of Control Processes*, Academic Press, Vol. I 1967, Vol. II 1971.
3. M. Athans and P. L. Falb, *Optimal Control*, McGraw-Hill,
1966.

[5] 1. D. L. Luenberger, *Optimization by Vector Space Methods*,
John Wiley, 1969.
2. D. L. Luenberger, *Introduction to Dynamic Systems*, John
Wiley, 1979.

AUTHOR INDEX

Abramovitz, M., 169
Anderson, B. D. O., 305
Arvidson, B. E., 15
Athans, M., 306
Auguston, R. M., 14
Bar-Ilan, A., 14, 83
Barrett, P. B., 16, 166
Bellman, R., 169, 305, 306
Ben-Haim, A., 14
Ben-Haim, Y., 80, 81, 166, 256, 296, 306
Berger, C. O., 167
Berger, J. O., 296
Betel, D., 15
Bevan, R., 168
Black, R. C., 16
Bondar, L., 83
Boster, T. A., 168
Boyce, I. S., 168
Brain, J. D., 167
Bram, J., 305
Bristow, Q., 226
Cameron, J. F., 168
Cejnar, F., 79
Chagaray, J., 296
Chernoff, H., 256
Chrysochoides, N., 168
Claisse, F., 15
Clayton, C. G., 227
Coderld, A., 15
Cohn, D. L., 225
Cohn, S. H., 15
Coop, K. L., 256
Costello, D. G., 80

Cottral, M. F., 15
Cowart, J. B., 226
Craig, A. T., 256
Cramer, H., 255
Cummingham, J. B., 168
Dell'Oro, P., 84
Despres, M., 83
Dickerson, M. H., 167
Dirac, P. A. M., 166
Diu, C. K., 167
Doob, J. L., 225
Dragnev, T., 83
Dudley, R. A., 14
Elias, E., 15, 168
Falb, P. L., 306
Fanger, H.-U., 15
Fehlau, P. E., 256
Feller, W. F., 79, 82
Fisher, R. A., 255
Fomin, S. V., 80
Foster, K. T., 167
Friedman, A., 167
Fry, F. A., 15, 224
Gardner, R. P., 15
Gettings, J. F., 83
Goans, R. E., 14, 167, 224
Good, W. M., 14, 224
Gozani, T., 15, 45, 79, 80, 81, 168
Green, B. M. R., 224
Greene, R. T., 167
Gudiksen, R. H., 167
Gunnick, R., 82
Hardy, G. H., 225

307

SUBJECT INDEX

311